CENTAUR

CENTAUR

Declan Murphy and Ami Rao

BLACK SWAN

TRANSWORLD PUBLISHERS
61–63 Uxbridge Road, London W5 5SA
www.penguin.co.uk

Transworld is part of the Penguin Random House group of companies
whose addresses can be found at global.penguinrandomhouse.com

First published in Great Britain in 2017 by Doubleday
an imprint of Transworld Publishers
Black Swan edition published 2018

A CIP catalogue record for this book
is available from the British Library.

ISBN
9781784162160

Typeset in 11.5/14.5pt Dante by Falcon Oast Graphic Art Ltd.
Printed and bound by Clays Ltd, Bungay, Suffolk.

To You

For You

Forever You

'If you're going to try, go all the way. Otherwise, don't even start. This could mean losing girlfriends, wives, relatives and maybe even your mind. It could mean not eating for three or four days. It could mean freezing on a park bench. It could mean jail. It could mean derision. It could mean mockery – isolation. Isolation is the gift. All the others are a test of your endurance, of how much you really want to do it. And, you'll do it, despite rejection and the worst odds. And it will be better than anything else you can imagine. If you're going to try, go all the way. There is no other feeling like that. You will be alone with the gods, and the nights will flame with fire. You will ride life straight to perfect laughter. It's the only good fight there is.'

Charles Bukowski, *Factotum*

Contents

Mayday

There is symphony in the movement of a horse.

The gallop, for example, is a four-beat rhythm: hind leg, hind leg, fore leg, fore leg.

You just have to listen for it, to hear it as I hear it, and you will realize how musical it is; how beautifully poetic.

This is the gait of the racehorse; it strikes off with its non-leading hind leg, then the inside hind foot hits the ground before the outside fore, but just by a split second. The movement concludes with the striking off of the leading leg, followed by a moment of suspension when – in glorious majesty – all four hooves are off the ground. Even at 35 or 40 mph, when the animal appears to be flying, it follows this classic, controlled cadence. In truth, it is not flying at all; it is dancing.

Hind leg, hind leg, fore leg, fore leg.

I can hum it in my head.

I have always followed this beat when I ride, moulding my body to the rhythm of my horse's stride pattern. And in this way, we have understood each other, my horse and I, our bodies in perfect sync, the energy between us reverberating like the silent echoes of an unspoken voice.

This is how I have always ridden. By an instinct, deep and wonderful.

It never failed me. Until the day it did.

May Day – Monday, 2 May 1994 – was a typical spring day at Haydock Park. The sun was shining brightly through a cloudless azure sky, the stands were packed with holiday-makers looking to have a grand day out. A gentle breeze blew across the racecourse, carrying happy voices, the tinkling of glasses, the familiar, very particular, scent of the horses . . .

Ominous feelings seemed improbable in an atmosphere like that. And yet, I was troubled. Just the day before, Ayrton Senna had died in a fatal crash at the San Marino Grand Prix.

I was haunted by this, by the emotions swimming around inside my head like demons. I couldn't quite compute them. At the most simplistic level, a life had been lost. That in itself was profoundly tragic. But it was more than that. Senna was no ordinary man. Three-time Formula One World Championship winner, he was considered by many to be the single greatest racing driver of the modern era. And yet, he had died.

What was more ironic, adding to the layers of sadness and confusion I felt over Senna's passing, was that Senna himself had been under immense emotional pressure on the day he died, following the death of Formula One colleague Roland Ratzenberger just one day prior. It was later revealed that a furled Austrian flag was found in Senna's car, which he had presumably intended to raise in honour of Ratzenberger after the race. That was never to be.

It was little wonder that Senna's death had cast a pall over the entire sports world. After all, we are all conditioned to believe that our heroes are invincible.

To me, it was deeply unsettling, perhaps because for the first time in my life, it brought home the stark reality of my own mortality.

I don't say this lightly. I don't say it with ignorance and I certainly don't say it with arrogance. I am just telling you the truth.

Yes, of course I knew that, like Senna, I was in a dangerous sport. But if you asked me if I had ever thought about death, I could look you in the eye and honestly say I hadn't.

Not because I didn't think it possible. *Only* because I did.

My teammate was a 1,200-pound animal, whose will I attempted to control at 35 mph. So to say that my profession was fraught with danger would be an understatement. But I was acutely aware of this. I approached race-riding with intuition and intelligence, equally split; there was no room for fear in this equation. I am not saying that I was fearless. How could I be? I'm only human. But I couldn't afford to consider fear as part of my reality. Fearing fear would have been as good as quitting. So I conquered fear with belief. I rejected fear. I shunned it. And to the extent I could, I tried to keep it at bay.

That I had suffered fewer falls than most of my colleagues was no coincidence. I was deliberate, measured and tactical. I would ride, feeling my horse's rhythm, the beat of its movement; I would time my every stride, approaching my obstacles with almost academic precision.

But the horse is an animal. An intelligent, unpredictable animal; a combination that can be as exhilarating as it can be deadly. And not even the most controlled and skilled jockey can predict which one it will be and when. So while I knew, as all jockeys do, that broken collarbones were lucky, death sad but not impossible, I certainly didn't want to think about it.

But Senna, uncomfortably, had forced me to.

These were the thoughts in my head as Jim Hogan, my friend, driver and former European champion distance runner, drove me up on that beautiful spring morning to the racetrack at Haydock Park for the Crowther Homes Swinton Handicap Hurdle.

On arriving at the racecourse, as was customary, I went into the weighing room and sat down in my place. Next to me was Charlie Swan, Irish champion rider, my rival and my friend.

I must have seemed preoccupied – not the picture of quiet calm that I was best known to present before a race – because he asked me if I was all right.

'Can you believe Ayrton Senna has died?' I said to him.

And quite unlike the light-hearted banter that usually takes place in a jockeys' weighing room on any normal day, Charlie and I exchanged thoughts on Senna, his untimely death, and our own vulnerability. It helped to talk to a friend. By the end of our conversation, my mind was more at ease. I felt more relaxed, more myself, more ready to face the challenge that awaited me in a few minutes – my race.

I put on my colours – distinctive red silks – and I went out into the paddock. Arcot was my only ride on the day. There were eighteen horses running in the Swinton Hurdle that afternoon, and I was riding the favourite.

From the moment I got on the horse, I shut everything else out. This was my way – blocking out the world, distractions fading into the background, giving way to a laser-like focus. When I am in this state of 'flow', nothing else matters. And so, just like the blinkers on my horse, I had blinkers on my mind. Senna – for the duration of the race – was forgotten. My only objective was to win.

As I started the race, I slotted Arcot into a good position, quickly finding his cruising speed. My judgement told me that there was sufficient pace in the race, so I got him to settle into a comfortable rhythm and I felt he did this well.

After jumping the first five hurdles, I considered that the front runners were going a stride too fast, but I was quite content to be where I was for the moment. It was when jumping the last hurdle on the far side that I started to creep a little bit closer – still completely effortlessly, still completely in control. I felt I had everything in front of me covered, and I felt good.

I was third in line approaching the second last hurdle and I could see the two horses in front of me were picking up speed again. I let out the rein, just a touch, maybe half an inch, and in doing so I was communicating to Arcot that I wanted him to lengthen his stride.

He didn't respond.

And with this, it all started to come undone.

I realized at this point that to have any chance of winning I would need two exceptional jumps from my horse, at the second last and last hurdles respectively. He came through impeccably at the second last, but I knew it was imperative for us to be in sync – body to body, muscle to muscle – to make that last and crucial jump.

There are a lot of things about race-riding that can't be learnt, they have to be felt. A jockey has to feel every movement of the horse underneath him and understand what each one means.

And going into that last hurdle, I felt something very telling from the movement of that horse beneath me. I realized three things, each separate but inextricably linked. First, that the stride ahead of us was too long. Second, that Arcot would have to *reach*, to extend himself, to make it. And third and most important, that in my judgement, I didn't think he could.

He just didn't have it in him – that *fight* – to make it.

So I made a decision that would alter the course of my life. I changed tactic.

Now, all that needed doing was for me to communicate my intent.

When I was a little boy, I had a Shetland pony called Roger. He was my pony and he was beautiful, small and lithe, a rich chestnut in colour, with a cheeky white face and two white legs. I loved Roger – he was intense and expressive, independent and strong-willed, with a definite rebellious streak within him. And, like me, he didn't like to be told what to do!

One chilly autumn day, I rode Roger out across the fields with the sole purpose of getting him to do what I wanted him to do. We started off well, jumping all the ditches across maybe three or four fields, Roger dutifully obeying my every command. I remember grinning from ear to ear, feeling smugly pleased with myself. Alas, a little too soon!

Because a few minutes later, when *I* decided it was time to turn back, Roger decided that *he* didn't want to. Try as I may, he wasn't going to jump the ditch on the way back.

You see, here is the thing:

Horses are living beings. They are spirited. They are stubborn. They are strong. In their natural state, they have a soul that is wild and free until the point they earn our trust. But even then, they have an instinct of their own. And the right to exercise it, at will.

So I tried and I tried, first by coaxing and cajoling, and then – as I grew more impatient – by chiding and rebuking. My efforts were completely futile. Roger wouldn't move.

Then, stranded three fields from home with no way of getting back, I had an idea.

I took my coat off and put it over Roger's head. His head completely covered, and unable to see where he was going, he had no choice but to place his trust in me. He quietened immediately and I reversed him slowly into the ditch, then I climbed on his back, pulled the coat off and waited. Finding himself in the ditch now, Roger had no option but to jump out. No sooner had he done that, I turned him around in the direction of home. Roger jumped the ditch and jumped all the way home.

Roger remained a pony of tremendous character, but something subtle changed from that day forth. Roger understood my will as much as he thought that he could force his will on me.

We had learnt to communicate.

At that last hurdle at Haydock Park, when I realized that my horse simply did not have the reserves to take off the way I had tactically imagined, I made a split-second decision.

I decided to give him the opportunity to take an extra stride before he made his jump. *But he didn't take it.*

Instead, Arcot decided to go for the hurdle. *Mayday! Mayday! Mayday!*

If I'd had the opportunity to take my coat off and put it over his head, so he could trust in my instinct, I would have grasped it in a heartbeat. But of course, I didn't. Instead, I stood by helplessly – stranded like that day three fields from home – while our communication, that crucial silent dialogue which bonds horse and rider, shattered like a thin, fragile sheet of glass.

And at that very instant – that fleeting snapshot in time – we rewrote my destiny.

In reality, the series of events that followed unfolded within milliseconds of each other, but in my head they seemed to play out in slow motion, frame by frame until time itself stood frozen in disbelief.

Arcot launched himself towards the belly of the hurdle in a valiant gesture of misplaced ambition.

Aboard him, I realized what he had just done. *My God, he's gone for it.*

But it was much more complicated than that.

The hyperextension of his body in reaching for the hurdle – the freakish, unnatural motion of it – snapped his pelvis in mid-flight, dragging his now lame rear end down to the ground.

My stomach turned. *He's not going to make it.*

I could sense he was frightened now, fighting me.

Then, fractions of a second later – as the same cold

horror of realization dawned on the beast – his front legs paddled helplessly in the air before they smashed into the timber of the frame.

Lame leg, lame leg, fore leg, fore leg.

There was a deafening crash.

And right then, in that moment of truth, I felt a strange calm, and my body and mind seemed somehow to disconnect, as if what happened next wasn't happening to me at all.

Just as my head flew forward with the momentum of the stride, the distressed horse's head flew back, our two skulls colliding with a sickening thud that pulsated through the air. I lost consciousness before I was catapulted off the horse and thrown to the ground. And there we were, man and beast lying motionless in one mangled, burning heap of broken flesh.

But it wasn't over yet.

In the quiet stillness of the moment came the deafening roar of the thunder of hooves. I was now lying unconscious, directly in the path of the horses approaching me from behind. Next in line to finish the race was Cockney Lad, carrying aboard him an ashen-faced Charlie Swan.

Trapped in the rhythm of his own beat, Cockney Lad advanced towards us, panic and fear filling his head when he saw our lifeless, tangled bodies on the ground ahead.

Two obstacles, one decision. Instinct screamed within him.

Frantically trying to avoid the larger of the two objects that lay stricken before him, he swerved away from the horse and *made his choice.*

Eyes wide in frenzied resignation, he thrust his body

forward and, capitulating under the force of his own momentum, his hoof hit my head, shattering my skull in twelve places.

He galloped over me in one final, feverish crescendo.

And then the music stopped.

Roger

The Aftermath

The dull, hollow thud of hoof hitting head – life hitting life – reverberated hideously through the air.

Then followed a deathly hush.

In a cloud of dust, the other horses galloped past me to finish the race that they had begun, the sound of their hooves drowning out the sound of silence.

Spectators gasped, colleagues prayed, reporters clocked the seconds. And everybody waited – frozen faces, bated breath – for me to awaken.

But I lay in a deep, silent sleep, while the blood gushed out of my mouth and on to the ground of Haydock Park, staining it a deep, vivid red.

My Ghost

My ghost belongs to a different world.

'Obviously!' you say. 'She's a *ghost*.'

'I don't want this to be a racing book,' I say.

'That's easy,' she says. 'I don't know anything about horses.'

I ask her what she knows of racing.

'Oh,' she says hopefully, 'I've been to Ladies Day at Ascot!'

I know then, that she's the one.

She makes me watch videos of my old races on YouTube – videos I didn't even know existed. 'You were a beautiful rider, Declan,' she says.

When I say nothing, she sighs. 'You mustn't take it lightly, you know. Your riding is brilliant – so natural, so effortless. You were meant to ride; some things are just meant to be.' Then, 'How could you not want to be a jockey and be so fucking good at it?'

'Don't swear,' I say. 'I don't swear. I want this to be an eloquent book.'

'OK,' she says after a touch of pause, 'I think I can manage eloquent.'

A moment later: 'Do you honestly not swear?'

'No,' I say. 'I don't swear, I didn't drink till I was thirty, I don't smoke, I don't gamble, I don't lie.'

She shakes her head then. 'I'm not writing your book unless you are perfectly honest with me, Declan Murphy. Everybody has sins.'

She's right, you know. The part about how some things are meant to be.

Her writing is my riding.

My Declan

They say lightning comes before thunder and thunder comes before the storm.

This is the natural order of things, nature's way of giving warning before disaster strikes.

On Monday, 2 May 1994, there was no such warning. The sky was blue, the sun shone bright. The thunder didn't come, and neither did the storm.

But the lightning did, even in the middle of the glorious sunshine.

And when that lightning struck, it struck with a vengeance. Little arcs of light danced across the roof of 5 Bank Place and then they all converged, to strike at its very heart.

At 1.45 p.m. on Monday, 2 May, my father, Tommy Murphy, walked through the kitchen of our home, at 5 Bank Place, where my mother was engrossed in baking her third batch of bread. He stopped for an imperceptible second, and then said casually, almost in passing, 'Mam, it's the last big race of the season. Declan's riding the favourite, I'm just going to put the television on.'

This was his way of asking my mother to watch the racing with him. He knew she had little interest in the sport, so he never asked explicitly.

Maura looked up from her dough and nodded. Every married couple has their own particular and idiosyncratic way of understanding one another. Tommy and Maura understood each other perfectly.

Five minutes later, at 1.50 p.m., Tommy settled himself comfortably into his favourite armchair and switched on the television. Tommy was content. Sitting in the chair made him happy – it occupied pride of place in the Murphy household, positioned strategically so that it was both directly in front of the TV as well as closest to the fireplace that bathed the room in warmth and light on many a chilly night.

Tommy was happy for another reason. There was nothing he enjoyed more than watching his son on TV, ghosting yet another winner past the finishing line, and into the annals of racing glory.

For the fact remained that Tommy Murphy was immensely proud of his youngest son. So much so, that the residents of the little town of Hospital (Irish: *An tOspidéal*, named after the crusading Knights Hospitaller) had become quite accustomed to the sight of him walking down the street, newspaper open, stopping to show people the photographs, as he declared in paternal delight, 'My Declan did this' or 'My Declan did that'.

To Tommy, the specifics of the 'this' or 'that' were of little relevance; the fact that it was 'My Declan' doing the 'this' or 'that' was what filled his heart with joy.

It was, My Declan.

Always, My Declan.

At 1.55 p.m., he focused his attention on the TV just as the jockeys began cantering down to the start. He could pick out his son straight away, even from a distance. There was something about the way he sat on a horse that made him easy to spot.

Just then, Maura Murphy walked into the sitting room, considerably less animated than her husband. She was a petite woman, with a kind face and a quiet, soft-spoken demeanour.

If anyone believed that watching three sons perform the same death-defying feat for a living had hardened her in any way, they couldn't be more wrong. When any of her boys were racing, she preferred not to watch, opting instead to go for a walk. When she did stay home, she watched each race while holding her breath, releasing it only with a 'Thanks be to God' when it was over. And so, as far as Maura was concerned, it mattered less if her sons won or lost as long as they were still on the horse's back when they finished the race.

Today was no different. She sat down reluctantly on the edge of the chair, hands together, fingers linked.

At 2 p.m. sharp, the Crowther Homes Swinton Handicap Hurdle began, as a field of eighteen ambitious, talented, seriously competitive jockeys battled for distinction.

As the race progressed, Tommy kept his eyes fixated on the figure in the red silks towards the back of the field, riding – in his characteristic style – a patient race. But Tommy knew he would pounce in the end, when the time was right. It was the way his son always rode, with his head – a clever rider. And so it was hard to contain the excitement in his voice when he spoke, 'He's going to win, my Declan is. He's going to win.'

'I've put the kettle on,' Maura said by way of reply, as if announcing the banality of the task would somehow cement her disinterest in her son's potential victory.

Precisely 3 minutes and 27 seconds into the race, on a racecourse 25 miles outside the city of Liverpool, a horse crashed into the last hurdle.

Even as the Irish sunbeams crept cheerfully across the room and the smell of freshly baked soda bread filled the air, a mother and a father watched their TV screen in horror as their son plummeted headfirst into the turf and then, within seconds, appeared to be kicked in the side of the head by an advancing horse.

Tommy gasped, a low, raspy noise coming from the bottom of his throat.

Maura got up and left the room without a word.

And the kettle started to boil.

Barney Curley

It all began with a man.

There are those we meet during the course of our lives who shape our destiny.

Sometimes, these are the people we are most conscious of.

Other times, they are just an idea.

I first met him on a Saturday night in the spring of 1984, sitting on the sofa of my living room. I was leading amateur jockey in Ireland at the time, all of seventeen years old. He was on the other side of my TV screen – tough, inscrutable, intelligent – star of *The Late Late Show.*

At this time, I was in my final year at school and every Saturday night, as a matter of course, I would go to the disco in either Galbally or Ballylanders or Kilballyowen with my friends John Farrell and Gerry Gallagher. Purely by chance, the night in question happened to be that rare one time when I didn't end up going out with my friends. And so, much against my adolescent will, I was subjected to *The Late Late Show* by my mother, who watched it religiously every Saturday night.

When you think about the fact that we had only two TV channels in the whole of the country at the time, RTÉ 1 and RTÉ 2, you will understand why *The Late Late Show* was such a phenomenon. The presenter, Gay Byrne, was a renowned, well-respected public figure, and whether or not it was intended this way, the show had become a forum where previously taboo social-interest topics were discussed openly, along with book reviews, music acts and guest visits from famous and interesting people – politicians, actors, authors and others with stories to tell.

Even then, he was a controversial figure – shrewd, outspoken and articulate, the perfect foil to the charismatic Gay Byrne – and though I had started off watching the show reluctantly, the dynamic duo had me hooked. He was being featured on the show because he had just successfully raffled his mansion in Mullingar and was talking about the gambles he had pulled off – some of the biggest betting coups in racing history.

I sat glued to the television, awestruck, filled with admiration and wonder. Here was a man who did his own thing, kept his own counsel. He had originally studied to become a Jesuit priest. Now he was one of the most famed people in racing, a legend of a punter. He had successfully outsmarted the system in ways that no one could even imagine, let alone execute, and all entirely legally. All of this struck a chord within me; I was totally, completely fascinated.

He went on to tell Gay Byrne that he had decided to move to England, and that he was taking all his horses with him. Then, he said – and my ears pricked up like a horse's – 'I'm going to take the best jockey that I consider in Ireland or England to ride them.'

Now, here's the thing about me, about my life, the fundamental truth that might surprise you, even shock you, and now's as good a time as any to tell you, and it is this: in all my life, I have never, ever harboured any ambition of making a career out of racing horses. There, I've said it. I didn't ever want to be a jockey, and this truth hasn't changed from the moment I got on my first horse to the moment I got off my last. It was a glorious hobby. I was riding for fun. And, for me, that's all there was to it.

But on hearing him make that remark on *The Late Late Show*, it sparked something within me. I remember saying to my mother, without even thinking about it, the words tumbling out almost as an automatic reflex, 'If I were ever to ride horses, I'd love to ride them for him.'

Why would I say this, you ask, a young boy of seventeen who had no interest in becoming a career jockey? What was it about this man that drew me to him in this way?

I don't know, is my honest answer. How can you say what's in a man? Only that whatever there was in him, I was fascinated by it. It was as simple as that. If he sold chickens, I would have wanted to sell chickens for him. It so happened that he needed someone to ride horses, and I could ride horses. And I could ride them as good as anyone else.

And so it began with a man.

A man who was unusual. A man who was ingenious. A man who was cleverer than anyone I could think of. A man who was like nobody I'd met before, like a character out of a novel or a movie – only he was real.

A man who changed my life.

That man was Barney Curley.

I was born Declan Joseph Murphy on 5 March 1966, to Tommy Murphy and Maura Murphy (née McCann).

The second youngest of eight, I was a precocious child, eager to learn, eager to absorb, ever-ready for a challenge, but always very much in control. There was a self-assurance about me even at a young age, a quiet confidence that somehow gave the impression that I was older than my years. Much later, in an interview for *Reader's Digest*, my father would say of me, 'When Declan was eight years old, he spoke like an adult.'

As a child, I was intuitive and self-aware; I knew what I

could do and what I couldn't, but when I believed I could do something, I generally did it. I was sensitive and shy on the inside, but outwardly, I came across as fearless, unfazed by it all.

I believe that the nucleus of who we are, as people, rarely ever changes. Because as much as I am a product of the experiences that have shaped me, I still believe that at the core, who I am today, my way of being, has always stayed true to the child I once was. I am still a deeply instinctive person, whether in my interactions with people or horses; my first impressions are rarely wrong. Trust and loyalty are sacrosanct to me, and when I offer someone my friendship, it usually lasts a lifetime.

This is the way I approached horses, it is the way I approached racing, and it is the way I approach life. It is the only way I know.

Growing up in rural Limerick amidst green, open spaces, horses were a natural part of my childhood. My father, an engineer for the local waterworks, was an avid follower of the sport but never rode himself. My Uncle Mikey did, however, as did most of my brothers and my sister Kathleen, who would end up being as good as, if not better than, the rest of us.

Horses and ponies were part of the family – how many we had at any given time depended largely on how much space we had to look after them. But my childhood wouldn't have been the same without them: Barney, Roger, Misty, Lightning – my first ponies, my first teachers and my first friends.

My earliest proper recollection of horses goes back to Strawberry, the very first pony my family owned. If you were to think of a white canvas, flecked all over with red-brown strawberry shapes, and stretched over the body of a horse, that was the very aptly named Strawberry for you. He was a big, strong-willed character, who had been an excellent show-jumping pony in his early life. We used him as a workhorse – my older brothers Laurence and Pat used to ride him and he would tow the cart to deliver firewood all around the village.

I had probably just started walking when I first became aware of Strawberry, but it is amazing how much of some things one can remember (and how little of others), and he stands out in my mind as the first introduction my big blue toddler eyes had to the wide world of equestrian wonderment.

It was maybe only a couple of years later, when I was about four years old, that I scrambled aboard my first pony to ride for fun. And what fun it was! My older brothers and sisters were already riding by now, so it was almost a rite of passage that I would join them on horseback, as soon as I possibly could. The curious thing with us, even as we grew older, is that we were never taught to ride, we learnt to do it ourselves. It was a normal part of everyday life; we rode horses like kids in other parts of the world ride bicycles – we would ride for pleasure, for the sheer joy of it.

Kathleen and I used to have races around the field, imagining we were riding in the Grand National, and our ponies were Red Rum and Ben Nevis. Or sometimes they were L'Escargot and Little Wolf – every horse that was on the television at the time, they were them! And when the Dublin Horse Show was on, we would pretend to be Eddie Macken and Paul Darragh, representing Ireland in show-jumping. We would hop up on two ponies, bareback and no bridles, and they would gallop down to the end of the field under the trees, where one of us would fall off and the other would be declared the winner!

Later, when I moved on to more sophisticated riding and was first shown how to use a saddle, I actually remember not understanding why people used saddles at all – it seemed so much more natural just to ride bareback. That was the way we all rode, starting out – no saddle, no bridle, just a head collar with a piece of rope tied to each side. It was mad, but that was the fun of it for us.

Looking back, I realize just how much I learnt from those early days of riding ponies – it really served to hone my natural instinct for handling horses. Sometimes, when you don't have the luxury of formal instruction, you don't overthink things, you just rely on 'feel'. All the time, you're talking to the pony, you're working it out between yourselves – how to stay on, how to get him to keep you on; the unspoken communication. This would prove such a valuable lesson later on in my life when I was out there riding racehorses at big races – it didn't matter really, the enormity of the event, the key to success was exactly the same. Instinct, intuition, judgement. Back to basics.

Once, when I was ten years old, my father took me to the horse races at Tipperary Racecourse, which is just a few miles from where we lived. My second-oldest brother, Pat, was already a jockey by this time, and my other brother Eamon was successfully racing ponies. At the track we met pony trainer Christy Doherty and, mistaking me for my brother Eamon, he looked at me and to my surprise and delight said, 'Eamon, would you ride my pony on

Sunday?' My father had stepped away for a minute, and I thought it the funniest thing in the world to pretend to be my brother, so I shrugged and I grinned and I said, 'Sure!'

On that Sunday, my father dropped me off at Newcastle West, where I jumped in the car with Christy Doherty and, with the pony, String of Pearls, safely in the trailer behind us, we set off for County Cork. It was when we were about halfway through our two-hour journey that the trainer stopped the car at the side of the road – there was nothing around us for as far as the eye could see – next to a gate that opened out on to a field. He then led the pony out of the trailer, took her into the field and let her have a long roll in the grass! He explained to me that a roll in the grass was as good as a stone of oats – it spoke volumes about the wellbeing of a horse. It was at this point, while String of Pearls was having a good old roll in a field in the middle of nowhere, that Christy Doherty started to realize I wasn't Eamon. He stopped mid-conversation, leaned forward, squinted at my face and asked, 'Are you Eamon?'

'No,' I replied truthfully, 'I'm Declan.'

He told me then that he was very sorry but he couldn't risk letting me ride his pony because of course I had never ridden competitively before. I was disappointed, naturally – I thought I would have really enjoyed this race – but I said, 'Sure, no problem, could I come along anyway?'

When we arrived at the races, I wandered around, taking in the atmosphere – the buzz, the banter, the boisterous laughter. It felt so vibrant, so electrifying. I wished I was participating in the excitement rather than merely spectating, but by now Christy Doherty had already found a more experienced rider for his pony. As luck would have it, however, I overheard Bosco McMahon at one of the horseboxes saying that they needed a rider for a pony in the 14.2hh race – the same race that I'd been hoping to ride as Eamon. So I pulled on the most earnest face I could conjure up, told them I'd ridden many winners before, when of course I'd never raced before in my life, and got myself on the pony.

Whether it was skill, fate, or plain dumb luck, I will never know, but there I was, racing my first pony, The Rake, in wellingtons and a borrowed helmet, and I won, beating the pony that I was taken to the races to ride. My win surprised trainers and the other competitors, who couldn't believe that a ten-year-old child who had never raced before had actually walked off with the prize. For me, it was my first taste of victory, and I was on top of the world.

The only thing sweeter than winning the race was the look on Christy Doherty's face as it happened; that still makes me laugh. He could never say my name properly, he always called me 'Del-can'. He walked up to me that day, after the event, and said, 'Oh, Del-can, you should have rode my pony!' 'It's Declan,' I remember saying to him, smiling innocently, 'not Del-can, but Declan.' And from then on and for evermore, in Christy Doherty's eyes, I went from being 'Del-can, the boy who had never ridden before' to 'Declan, the boy who could ride'.

Suddenly, a whole new world had opened up for me. Within that year I became champion pony rider, going on to retain the title for three years. By the age of fourteen, I had ridden a couple of hundred winners. But as much as I enjoyed winning, one thing stayed constant throughout.

I was riding because I enjoyed it, not because I wanted to make a life from it.

No, I wanted to be a lawyer. I wanted to live in New York. I wanted to see the world. Every Thursday morning, without fail, I used to read Jenny Bannister's 'American Diary' in the *Irish Independent* – I had eyes for America long before I ever set foot there. I had always been considered well-spoken and articulate, choosing my words carefully and delivering them with impact. So it wasn't an impossible dream that I saw myself in America: Declan Murphy, criminal lawyer, dressed in a tailor-made suit and shiny new shoes, defending my clients in courtrooms filled with people. I didn't dream to live; I lived to dream. I lived my life, but I dreamt of something else, of something beyond the life I had.

My mother, Maura Murphy, dreamt my dream. While my father was always excited when I raced, and more so when I won, my mother never wanted me to ride horses, least of all as my profession. While part of this was driven by some larger ambition for me, another part – consciously or subconsciously – almost certainly had to do with the risk involved with being a jockey. I know she found no logic in placing one's life at the mercy of half a tonne of pure muscle, for no reason other than to satisfy that aberrant thrill-seeking gene which some of us are apparently born with.

A simple girl from Portarlington, she had met my father when he was working up there as an engineer for Bord na Móna. They had fallen in love, and moved back to Hospital to make a life together. Most of the McCanns – her elder sister Noeline, her brother Michael, her brother

Bobby – had all emigrated to America right after leaving school, and she believed I was destined to do something bigger with my life. 'You will do something extraordinary,' she would say to me from time to time, 'I just know it.'

This was her dream. And, in my heart, mine.

Meanwhile, back in the real world . . .

My dexterity with ponies was beginning to attract attention and for the first time, people started making reference to 'my hands'. With horses, the reference to hands is metaphorical – it is an all-encompassing term that connotes something much more intangible. When people said I had a good pair of hands, they were alluding to a rider's ability to 'feel' his horse – horses *need* that feel, they need that confidence you emit and transmit to them. You bounce it off each other. I had this even from those early years – that 'feel' in 'my hands'. Horses seemed to respond to it, to relax to it, to settle so naturally and beautifully because of it. And this, for me, was the reason I continued to ride and to race; on some level, it seemed that it was ingrained in me. You are either born with it or not, and I was learning slowly that, in some capacity or another, I was destined to ride. Where I chose to take it was another matter altogether.

And so, from around the age of ten to the age of fourteen, I continued pony racing with great gusto. I realized it wasn't something that I had to try too hard to be good at; it came naturally to me and, as a teenage boy, that suited me brilliantly: the less I had to work at something, the more I tended to enjoy it!

Nineteen eighty was the last year I was champion pony rider. I was fourteen years old then and I could no longer do 8 stone; I had just grown too heavy for pony racing. I started to take a keen interest in hunting, a strong Irish tradition and a natural hobby for many riding enthusiasts.

Fortuitously for me, exactly around this time, 5 Bank Place welcomed Misty into its equine family. Misty was a 14.2hh pony that we bought from Dan Donovan, a friend of my father's, who lived in the neighbouring village of Kilteely. Dan was leaving home to work with horses on the Curragh and so, much to our delight, Misty became ours.

There are jumpers and then there are jumpers, and Misty was an unbelievable jumper. She was mercurial by nature, but she never ceased to amaze with her ability, and we would watch wonderstruck as she jumped double ditches with the agility of a cat. For me, Misty couldn't have arrived at a better time – she became the perfect companion to ride hunting.

At the time in Ireland, hunting was largely synonymous with the Ryan family; Thady Ryan was Master of the Scarteen Hounds, Ireland's most famous pack of foxhounds. The Ryans had run the Scarteen Hunt – named after their house in Knocklong, on the Limerick–Tipperary border – since the late eighteenth century, and their distinctive Black and Tan Hounds had since become a celebrated feature of the Irish countryside.

Misty and I eagerly joined the hunt – and it wasn't long before we developed quite the reputation. A typical hunting party follows a strict hierarchy: leading the front is the master huntsman, followed by the whipper-in. After that, where you sit is determined by relative authority, and kids like me were meant to follow at the very rear. But Misty was such a brilliant jumper that it was easy to get carried away. Often, I'd find myself at the front with all the adults, swept away by my own enthusiasm, inadvertently breaking rank.

Thady Ryan and his whipper-in, Tommy O'Dwyer, were amused no end by my antics – they viewed me as an audacious little kid with both the ability and the confidence to negotiate the rules. But I was often told off by the other huntspeople for my disobedience! I had always had a mischievous side, and a part of me enjoyed challenging authority. Even at fourteen, I was driven by a cool self-confidence – I knew that even though I was young, I could hold my own on horseback. And so in my own head, I wasn't breaking any rules, only reinventing them!

It wasn't until much later that I realized how much of an education hunting gave me in preparation for being a jump jockey. Riding over hurdles or fences is a tremendous skill, but you have the benefit of knowing where the obstacles lie – you see them, so you prepare for them. You calculate, you strategize, you plan. With hunting, you're riding virtually blind. You don't know the lay of the land, obstacles come up unexpectedly and you've got to have the confidence to jump them as you see them, without the benefit of foresight. I found that hunting lent finesse to my riding, it sharpened my instinct for confronting obstacles, it forced me to make split-second decisions under pressure and, most important of all, it unequivocally tested the limits of my courage.

It would be remiss at this time not to mention the indispensable part Misty played in honing my skill as a rider. Not only was she a star jumper, but her temperament was such that it pushed me to fine-tune my own prowess. She was so headstrong, so erratic when she was in the gallop or while jumping, that she consistently challenged my ability to ride her. Misty was a creature that, for some reason, wanted so much to be out of control. This forced me to raise my game just so that I could keep her *in* control.

All of these individual experiences – my early exposure to horses, the hunting, the jumping, the pony racing, Misty – formed the vital ingredients that had unsuspectingly been thrown together inside the cauldron of my mind. And now the potion was brewing within me, creating something weird and wonderful, something that would become apparent to others long before I became aware of it myself, in a form that would be widely regarded as artistry. And yet in its infancy – in that delicate embryonic stage – it was no more than a young boy trying to control the animal he was riding.

And so it followed.

Until one day . . .

I was out in the early hours of the morning, hunting on Misty. We were on a steep, downward-sloping dirt path in the woods when I turned her around to jump a five-bar gate. It was either bad luck or an error of judgement, or perhaps a bit of both, but Misty caught the top of the gate, threw me over, and then fell with her full weight squarely on one of my legs. What followed was a visit to the hospital, where I was told in no uncertain terms that my leg had been badly injured and needed complete rest. My mother – rarely one to lose her cool – lost her cool. That put an end to hunting for a short while.

It was around this time that my father bought me Tarenaga, a 15.1hh horse, from Tommy Kinane. Albeit small, Tarenaga was a racehorse – a proper racehorse – who I got to ride at speed. This was when the addiction really started. I began to love the thrill of racing, of doing it right, of perfecting the ability to coordinate balance and pace and technique to achieve a race-rider's one ideal: speed.

When I was sixteen, Eamon, by then an apprentice jockey on the Curragh, invited me up to join him for the summer, where he got me a job with Kevin Prendergast, one of Ireland's consistently successful trainers. I was to ride his horses to exercise them, for the three months of my summer holidays. One morning, when I was riding a notoriously headstrong horse, Piccadilly Lord, I noticed Kevin Prendergast watching me handle him, a look of admiration in his eyes. The Curragh is filled with racehorses, filled with people involved with horses, but more than that, I went to a stable which, at that moment in time, probably had about ten seriously good aspiring young jockeys. To be able to shine through bright enough to make an impression in that particular environment in the yard, on the gallops, perhaps spoke to my ability, except I didn't see it that way. I was just a kid playing around with horses in his school holidays.

The next morning, Kevin Prendergast came up and spoke to me. It was the first time anyone had seriously broached the subject of my career. He said, 'I like the way you ride – you have natural talent. You should be a jockey.'

I said, 'Thank you, but I'm going back to school, I have no interest in being a jockey.'

And so, I went back to school.

The following summer – it was 1983 and I was seventeen

years old now – I went back to ride at Kevin Prendergast's for the holidays. One day, he rode up next to me and said in all sincerity, 'Would you like to be an amateur jockey?'

Matching sincerity for sincerity, I said, 'No, not really.'

He said, 'Why don't you take out an amateur licence? It won't cost you anything.'

I said, 'But I don't really want it. I'm going back to school in September and I won't be able to ride.'

And so he let the matter drop . . .

But Kevin Prendergast was the leading trainer at the time in producing the next generation of apprentice jockeys, and 'talent spotting' came naturally to him. It seemed he wasn't used to taking no for an answer.

A week later he rode up next to me again – I remember it was a Thursday morning – and he said, 'I've got an interview for you tomorrow with the stewards at Navan Racecourse.'

I said, 'An interview for what?'

He said, 'For your amateur licence.'

I said, 'I don't want an amateur licence.'

He said, 'It's only a day out for you. Just go for the interview, it won't cost you anything and you can have a ride on Monday.'

The 'ride on Monday' happened to be the Friends Handicap, the feature event at the Rose of Tralee Festival. So, on Kevin Prendergast's insistence, I rocked up to ride my horse, Prom, 20/1 in this big amateur handicap, never having ridden in a proper race before.

I was riding against some leading figures at that time – Ted Walsh and Willie Mullins – but I was completely undaunted by their celebrity. In fact, I remember thinking to myself, What a lot of fun this is going to be!

But I was also there to win.

Winning is like a predator's first taste of blood – it hooks you. Notwithstanding the fact that I viewed racing as my 'recreational buzz', I was realizing that I had become competitive. Some of us thrive under pressure, while others of us crumble beneath it. I found myself driven by the pressure to perform, by the hunger to win.

The race was every bit as thrilling as I had expected. There were thirteen runners and I deliberately rode a waiting race on the hard-pulling Prom, getting him to settle beautifully. Turning into the straight, I still had a lot of horses in front of me and I started to make my move up the outside when I caught sight of the furlong pole. As soon as I saw it, I squeezed up my mount and his response was immediate. I won the race by four lengths.

Suddenly, I was two out of two. My first-ever ride pony racing, I had won. My first-ever ride as an amateur jockey, I had won.

The statistics became significant to everyone around me, except myself.

Don't get me wrong, I had a love of horses and a love of riding, a love of competition and a love of winning. Put it all together and you arrive at a single inevitable outcome. But I didn't see it that way. Winning these races was the equivalent, for me, to teenagers joyriding in cars. It was fun. And it was still just that – a glorious hobby, a bit of fun.

Having fun defined my formative years. As a person, I was easy-going and happy-go-lucky – a typical teenage boy. I thrived on an insatiable curiosity; I enjoyed pushing the boundaries.

When I was fifteen or sixteen, my sister Kathleen and I would routinely thumb lifts to discos in Knocklong and at the Bruff Rugby Club. One night, after everyone was sound asleep – it was raining the way it only ever rains in Ireland: that distinctive soft, steady drizzle – I decided that driving ourselves to the disco would be more fun than thumbing a lift in the rain. So I told Kathleen to go inside the house and get the spare keys to the car. Keys safely in my possession, Kathleen, I and our friend Gerry Gallagher pushed the car out on to the road.

I then drove us to the disco at the Bruff Rugby Club and we all had a grand night out. The following morning, straight-faced and as sincere as sunlight, I told our mother that it was Kathleen who had taken the car and that I had only gone along – with great reluctance – to make sure nothing happened to her or the car! Kathleen never got into trouble, so it's more than likely that our mother didn't quite buy my solemn impression of a concerned big brother, but she never said a word!

I had the same carefree approach with riding – I never took it too seriously. The day my sister Geraldine got married, I had been asked to ride a horse called The Yank Brown for Edwina Finn and Mickie Lee. I was doing my Leaving Certificate at school at the same time. So I did an exam in the morning, attended the wedding in the afternoon, and just when we got to the reception, which was in the Glen of Aherlow, I dashed off to Tipperary Racecourse, where I proceeded to ride the race, win the race, and come back to the reception before they had all even sat down!

And in this way I would have continued, riding only to

fuel what I thought was a passing interest during a temporary phase in my life, had fate not intervened.

When I had just turned eighteen years old, I was out riding my pony on the road one day when I was approached by John Power, solicitor to Barney Curley. He stopped me and said, 'Barney Curley would like to meet you.'

I said, 'Who's Barney Curley?'

And then I remembered that he was the guy from *The Late Late Show.*

'Oh, sure,' I said. 'I'll meet him.'

When I told Kevin Prendergast later that Barney Curley wanted to meet me, he said, 'Don't go near him. *Go* near him – find out what he wants – but don't do anything else. He will destroy your life.'

Words have power; tremendous power.

Especially words of warning.

Especially when spoken by someone you respect.

Especially when you are eighteen.

Words have the power to stop you.

They didn't stop me.

In the summer of 1984, my brother Eamon and his friend Mull took me to meet Barney Curley at his mansion in Mullingar – the very one he had been talking about on the TV show – and we sat, facing each other, on two chairs in the archway of this now infamous mansion.

Serendipity. I think of that word, its substance, and I take pause.

From my point of view, I was there because Barney Curley wanted to meet me. The idea of riding horses or becoming a jockey couldn't have been further from my mind. No, I was there to meet him because *he* wanted to meet *me*. And yes, of course, ever since I'd seen him on *The Late Late Show*, I had wanted to meet him too. He was even more impressive in real life than he was on TV – arrogant, detached, supremely confident. I guess there was an element of mystique about him, about this shaven-headed, cigarette-smoking rebel with the icy nonchalance and the perfect pencil moustache. I was completely drawn in. I struggled to get my head around how somebody so calm and undemonstrative in nature could go about doing things the way he did. This was a man who was doing everything himself in his own way of thinking, the sort of person you meet once in a lifetime. And even though I was only eighteen years old, I was inspired by this independence of mind, by someone who didn't need to seek counsel on decisions he chose to make.

I had expected our meeting to last ten or fifteen minutes but three hours later, my life was in a different place. You see, the thing about Barney Curley is that there is no one who holds his cards closer to his chest. There's no way of knowing what he's thinking, he never gives anything away. He could have just won or lost two million pounds and you would never know. Because Barney Curley doesn't even make an expression. And yet, somehow, I had managed to impress this man who couldn't ever be impressed.

This man, with his meticulous planning and uncommonly astute judgement, seemed as interested in me as I was in him. I told you I have always been bold and spirited, a precocious boy. And perhaps it was this innate tenacity that gave me the confidence to go and sit with someone like Barney Curley and hold my own. Later, in an interview, he would say about me, 'He was so calm and unintimidated and he spoke so well, I knew he was the one. I didn't even have to see him ride.'

Meanwhile, close to three hours into our meeting, I still had no idea why this man had wanted to meet me. And then, just as curiously as he entered my life, so he changed it. He said, 'I'm looking for a jockey to ride my horses in Newmarket. Do you think you'll be good enough?'

'Don't worry,' I said. 'I'll be good enough.'

Good Enough

The ability to ride a horse well is the ability to embrace a partnership that is unique.

It is a partnership between the horse and the rider in which the understanding is implicit – the horse is the true athlete; the rider is simply the enabler. And to enable the horse to run its best race, to enable the horse to win, is to master the act of balancing authority with kinship. The key to success is the ability to empower the horse to understand your will, while still allowing it to follow its natural competitive instinct.

So we communicate without words, through body language and movement, with an animal that survives in the wild on instinct alone. It is necessary to read the invisible signs. To learn the language of an animal that does not speak.

Race-riding as a sport has a language and a rhythm and an intensity that sets itself apart. I have always understood this language. This is, perhaps, what set me apart.

I always did things a little differently. My decision to come to England with Barney Curley, for example, was a decision reached without the consensus of the majority. In fact,

I asked exactly thirty people for their opinion and exactly twenty-nine of them told me in no uncertain terms that I was mad to even consider it. Kevin Prendergast had been unceremoniously blunt: 'You'll be back home within six months,' he said, 'and your boss'll be in jail.' But of course I did it anyway. Because nobody understood. I had no real interest in racing; I was just interested in the man. And the fact that I could ride meant that he was also interested in me.

But riding for Barney was not for the faint-hearted.

When Barney Curley voiced his opinion on something, he stood by it, and anything he voiced it over, he usually had plenty of knowledge about. The confidence, the ingenuity and the daring that so typified the self-acclaimed 'bandit of British racing' served to challenge me, allowing me to push the limits of my skill in ways that I may not otherwise have had the opportunity or the audacity to do. So, whatever my reasons might have been, I am certain that I couldn't have obtained the racing education that I received from Barney with anybody else.

One of the best examples of this was my experience riding Barney Curley's The Hacienderos at Kempton on Boxing Day 1984. Before the race, horse-racing pundit John McCririck had announced on Channel 4 television that 'the bookmakers were running scared like Christmas turkeys over Barney Curley's The Hacienderos', such was the flood of money that Barney had on the horse; it was a *huge* punt.

Interestingly, I had already made my acquaintance with the four-legged beast named The Hacienderos in a different life. It was in April of 1984. I was leading the amateur title while still at school, when Kevin Prendergast asked me if I wouldn't mind taking a day off for the Fairyhouse Festival to ride his horse, Conclusive, in the bumper.

Given I was busy studying for my Leaving Certificate at the time, I was reluctant to miss a day of school, but Kevin Prendergast was as charming and as compelling as only he could be. 'This horse will win,' he had said, conclusively.

So, of course, I went.

Before the start of the race, Kevin Prendergast and I were in the paddock, looking at the horses walking around the ring. He motioned at the horses one by one as they walked past, and said to me, 'Oh, we will beat this one, definitely beat that one, and that one, and that one.' But then he pointed to a horse – a big, beautiful bay gelding – and he said, 'But that horse there, that is one of Barney Curley's, you wouldn't know *what* that is, it could be anything! But, apart from that, we will win.'

So I went out and rode what I thought was a pretty good race on Conclusive, this classy, well-bred horse, and he looked like he was going to be very impressive. Turning into the straight, I'd just given him office to go, and he looked like he was going to win. And then Willie Mullins appeared, all guns blazing, aboard this horse that passed me like the wind with his tail out, and I thought, Where the hell did that come from? Sure enough, it was Barney Curley's The Hacienderos, an anti-establishment horse, owned by an anti-establishment man, a secret weapon, a beauty. I came back then, after getting beaten, and said to Kevin Prendergast, 'This was no reflection on Conclusive – that horse that came out of nowhere and beat us? I've never seen anything like that in my life! That horse was jet-propelled!'

I would never have guessed that just seven months later, in November 1984, I would witness The Hacienderos in action again. The race would be at Newbury on Hennessy Day, and once again, he would be racing across the finish line with his tail out, winning by two lengths. But this time, it would be with me on his back. It was my first time on the horse and one of my first winners as Barney Curley's jockey. I had expected Barney Curley to be unreservedly delighted, but Barney Curley was rarely, if ever, unreservedly delighted.

As I would soon find out.

A month after Newbury came the historic Boxing Day races at Kempton, and my second opportunity to prove my worth on The Hacienderos. Come race day, however, Barney Curley gave me a talking-to. On matters he considered weighty, Barney's tone was always such that, even when he didn't intend to, he sounded stern. And whether you are eighteen or eighty, when Barney speaks to you in that tone, you tend to take it seriously. 'You won too easy at Newbury,' he said in his characteristic dispassionate manner. 'You don't need to win by more than a length.' I remember being nothing short of amazed at his level of expectation, thinking that this degree of precision in judging the pace would have confounded the best of jockeys – it was a 2-mile race, how does one gauge the distance to a length? But, of course, a challenge such as this only served to fuel my fire. That was the brief that had been given, and that was the brief I was about to fulfil – the game had just become exciting.

The race went according to plan. Turning into the straight, I had plenty of horse underneath me and I could have gone around the horses in front of me, but I decided – valiantly – to go up the inside instead. Riding one of the horses ahead was seasoned jockey Steve Smith Eccles, who glared at me furiously, screaming, 'Where are you fucking going, kid?' as I rode past.

In his eyes, I had violated one of the unspoken commandments of horse racing: 'Thou dare not go up the inside' (unless you're certain you've got enough horse to get there).

He couldn't believe how such a young amateur jockey, fresh out of school, could exude the kind of brazen confidence that I was riding with. 'To the winning post!' I wanted to shout out, in answer to his question, but of course I'd raced by so quick, I was gone.

I was second in line jumping the second last hurdle and I knew then that I had everything behind me beat. Going into the last hurdle, I was seven lengths down, but I remembered what Barney Curley had asked of me, so I waited patiently. At this point, Graham Goode was saying on the commentary, 'I hope Mr D. Murphy realizes where the winning post is.'

I did.

Because exactly then – as casually as a stroll by the river on a summer's day – I unleashed the full might of The Haciendercos. I won by *a head*. Barney Curley had asked me to win by a length; I won by a head. I just couldn't help myself.

Later, Lord John Oaksey would say on Channel 4 television, 'Declan Murphy pulled the fat out of the fire, unnecessarily so in the first place. It was a grossly overconfident ride.'

Barney Curley would say that it was as good a ride as he had ever seen.

Steve Smith Eccles would come into the weighing room, fire his saddle on the table, look at me sitting there and say, 'You are fucking living dangerously, kid!'

And I would look up at him with a broad smile and say, 'But I queued up for your autograph at Limerick Races twelve months ago?'

This was my first proper introduction to the English racing public. Or, I suppose, it was the English racing public's first proper introduction to me.

It seemed, I was good enough.

These were the moments that made my head spin with the love of winning – these were the highs. But, equally, I had my fair share of lows.

It was certainly not all smooth sailing for me, and at times it seemed a very rocky passage to traverse. In my initial year with Barney Curley, there was talk of a witch hunt. The side effect of Barney Curley's reputation meant a constant clashing with the authorities, and I, by association, faced the wrath of the stewards on more than one occasion.

At Windsor, stewards grilled both Barney Curley's trainer, Dave Thom, and myself over the running of Experimenting in the Royal Borough Novices' Hurdle, where he finished fourth. Despite Thom's explanation that it was Experimenting's first appearance on a racecourse, and my own assessment that the horse had tired after the turn into the straight, but ran on better ground from the second last, the stewards claimed that the horse had been given 'no chance of winning'.

Within the space of four days, Barney Curley's integrity was questioned once again at Kempton when his horse, The Tariahs, trailed in eighth in the Motorway Novices' Hurdle. Thom, who subsequently found that the horse was sick and had to stay at the vet all night to get his stomach pumped, was incensed at the allegations levelled his way. Greater, however, seemed to be his concern for me. 'Declan is the best prospect I've seen for years,' he told the press. 'He's got a quiet style and I just hope that the stewards don't spoil the kid's future.'

But I remained unruffled despite it all. Every race I rode for Barney Curley, I was watched like a hawk, but I refused to allow these incidents to affect me in any way. Just like I had a job to do to the best of my ability, so the stewards had a job to do to the best of their ability. Some days I would ride a bad race; some days they would make a bad decision. I didn't take it personally – I knew I had ridden straight and honest races.

Notwithstanding the occasional setbacks I suffered simply by virtue of being Barney Curley's jockey, I had no complaints. If one door shut, another door opened – life usually had a way of working out. I focused less on the disappointments and more on the opportunities, and Barney Curley offered me many. For example, in the summer of my first year with him – and in the following four summers – I accompanied him to the west coast of America. The American Dream – as described so alluringly by Jenny Bannister in the *Irish Independent* – had beckoned me ever since I was a little boy, and Barney gave me the opportunity to chase my dream.

Going to America opened my eyes to a brave new world

of experiences. Here, I met Charlie Whittingham, one of the most celebrated trainers in US racing history, reverentially alluded to as 'the Bald Eagle'.

There are not many nineteen-year-old Irish boys who get lucky enough to meet Charlie Whittingham, let alone spend five entire summers learning from him; I recited the Lord's Prayer every night.

My time with Charlie was every bit as valuable as I imagined it to be. I had the opportunity to ride some top-class horses: Kentucky Derby winner Ferdinand, Greinton, and the talented Dahar, who will always be a part of my story. More importantly, I learnt a tremendous amount from Charlie himself – from his direct, understated way – that I would carry with me throughout my racing career.

Having just had experience of trainers in Ireland and England, one of the most striking attributes of Charlie Whittingham was his remarkable ability to simplify things. For example, I was riding a horse called Áras an Uachtaráin in his morning exercise one day. He was a horse that I knew well, he had been trained by Vincent O'Brien in Ireland, and I felt that Charlie had improved him, 're-moulded' him somehow – he seemed in such great form.

So I asked Charlie. I said, 'What have you done to improve this horse?'

And he thought about my question and he looked at me and walked alongside me, and he said, 'I've done nothing to improve him, I've just found a system that works best for him.'

It spoke volumes about Charlie Whittingham, about his sheer genius as a trainer. He was extraordinarily successful, but he was extraordinarily modest. He used his words sparingly, but whenever he spoke, he spoke with great impact. He was tough, but he was kind. He would tell you in his direct, candid style what was right and what was wrong. Under Charlie's eagle eye, I was really able to hone my talent by riding gallops against the clock. I surprised myself and sometimes I even managed to surprise him.

It was two days before the Fourth of July Sunset Handicap at Hollywood Park. I was riding Dahar, owned by Nelson Bunker Hunt, trained by Charlie Whittingham, in a workout, and all the clock-watchers were out, clocking the horses.

Hollywood Park racetrack was set up so you enter the track at the bottom of the home straight. Dahar and I jogged for the length of the straight, but I knew the horse was only going to breeze for 3 furlongs. We broke into canter, and I eased my way in at the 4-furlong pole, so that I was at full pace by the 3-furlong pole. Then I let Dahar run in my hands and quicken half a stride faster for the last furlong.

So magnificently had the natural competitiveness of the animal been sparked, that he was able to do this in 33 seconds and change. I thought his performance had been extraordinary but the clock-watchers were up in arms: 'He's gone too fast, he's burned the horse out,' they screamed.

Charlie Whittingham's assistant trainer, Rodney Rash, was a tall, handsome man, standing at 6 feet 3 inches, with broad shoulders, lean and taut and full of muscle. And somehow, he was totally oblivious to the magnetic effect he had on the ladies. I would often wonder at his cool, nonchalant ways, and during the course of my time in California, we would have many laughs together.

At this time, Rodney walked up to me and said, 'You motherfucker, what were you thinking? You pushed him way too hard.'

But the big boss never said a word.

Instead, Charlie Whittingham followed me until the grooms had taken the horse off me and I was standing in the barn alone. Then, he snuck up behind me, quiet as a mouse, and I heard his voice in my right ear, low and intense. 'Was that as good as it looked?' he asked.

I twirled around, being the exuberant kid that I was, and my voice brimming over with enthusiasm, I said, 'It was better! That horse could have done better!'

Charlie Whittingham walked away as quietly as he had appeared, without another word.

That Sunday, Dahar came second in the $287,500 Sunset, beaten by Zoffany by half a length. He had lost his race on the home turn when he got fanned wide at the precise moment in time that Zoffany, brilliantly ridden by Eddie Delahoussaye, snuck through a gap on the inside – tactics that had been thought out by his highly acclaimed trainer, John Gosden, prior to the race.

The groom that looked after Dahar was a towering African-American man, with a big booming voice, called Mr Ed. He always seemed very fond of me, but in all my time in California, I never understood a word he said – he didn't have a tooth in his mouth! On that Sunday, Mr Ed hugged me so hard I thought he was going to squash me, all the while mumbling something that will for evermore remain one of the most endearing mysteries of my life.

The next morning, on the Monday, Alex Solis, the jockey who rode Dahar in the race, came to the barn and handed me a cheque for $500.

But Charlie Whittingham still did not utter a word.

Later on, Barney Curley would ask Whittingham what advice the veteran could pass on, from one trainer to another. 'I expected him to say, feed this or feed that,' Barney recalled, 'but instead he said to me, "Keep the numbers small . . . and never lose that kid Murphy."'

It seemed, I was good enough.

Back in England, I continued to ride in the characteristically quiet style that would soon become my trademark.

I modelled myself on a combination of Steve Cauthen and John Francome – arguably two of the most iconic jockeys of all time.

I was still in school the first time I watched John Francome ride. I had never seen a horseman with such grace and elegance and poise. I remember standing next to a hurdle and watching his horse jump – I had never seen anybody clear an obstacle so artfully. From that moment on, I was a fan. When I came over to England, I tried to watch John Francome ride whenever I could and learn from him. He rode with such effortless ease; in my mind, I thought, Now *that* is a jockey.

It was the same with Steve Cauthen, the American jockey who rode for Sir Henry Cecil. He had a similar riding style to that of John Francome, just on the flat. Steve Cauthen on a horse was a thing of beauty – nothing appeared forced; his movements were so utterly graceful, they just seemed to flow.

They were both intelligent men, elegant riders who rode tactically and stylishly rather than relying on just strength. And while I was still young, trying to be the best I could be with what I had, I thought those two jockeys were the greatest examples of the best way to ride.

Many years later, I would be fortunate enough to be compared to both my role models. After the Tripleprint Gold Cup in 1993, when I successfully led Fragrant Dawn to victory, John Carter of the *Independent* would say, 'Murphy's nonchalant nursing of this doubtful stayer, bringing him through from almost the next parish to glide past the leader in the shadow of the post, was reminiscent of the former champion John Francome at his arrogant best.'

In January 1994, after I won an appeal against a two-day ban imposed on me by the Jockey Club, Monty Court, former editor of the *Sporting Life*, would say, 'Declan Murphy has been the most confident, articulate occupant of the weighing room since the day Steve Cauthen walked out of racing.'

It seemed, I would be good enough.

Slowly, as I began to establish my name on the riding circuits, my distinctive style began to be recognized, and my list of appreciative admirers grew by the day.

But in those early years, the trouble with the stewards didn't let up. The Jockey Club seemed determined to stick a knife into Barney Curley, but he was just too shrewd for them. Whatever they planned, the free-punting, shaven-headed gambler was always one step ahead. I proved a far easier target, so they decided instead to get at him through me.

They were relentless. It seemed at one stage that I couldn't take part in any race without being called before the stewards for something or other. All this time, I had remained unfazed, but the injustice of repeatedly being the proverbial goat of Leviticus was beginning to affect me adversely.

One time at Sedgefield, I was accused of throwing a race I didn't win because the stewards argued that I hadn't 'tried hard enough'.

Twelve days prior, I had broken my collarbone while riding at Towcester. Solvent was my comeback ride at the Road Show Novices' Hurdle at Sedgefield – my first race since the injury.

On the day, my collarbone felt healed. Happily, I had been passed fit to ride and, being the eager young jockey that I was, I was determined to do my best. Solvent was 4/1 second favourite and my instructions from Barney Curley had been to lie in about fourth or fifth position and make my effort after the second last.

So I had ridden out and my collarbone had been fine, without undue stress, but right from the start, I had been unable to keep Solvent well covered because the horse had been racing too freely. Then, when Solvent pecked at the first, I jolted my shoulder so badly that the bone cracked again.

The pain was so bad that my left arm became totally disabled from the fourth last. Despite my best efforts, we lost our place coming out of the back straight, falling to sixth by the last flight. It was not until the run-in that I used my right hand to compensate and we ended up finishing fourth, a length and three-quarters behind the winner.

To my shock, I was fined £150 on the spot by local stewards for breaching Rule 151 (ii) which deals with 'giving a horse a full opportunity to win or obtain the best possible placing'.

Adamant to prove my innocence, I went to the hospital on my way back from the races that very day and came home with a strapped-up arm and X-rays that clearly showed my collarbone had cracked again. In my mind, this proved beyond reasonable doubt that I had been in unbearable pain from the point in the race at which it broke.

But even medical evidence proved insufficient. In fact, to my astonishment, it only served to fan the flames. I was summoned to an inquiry by the Jockey Club's Disciplinary Committee at Portman Square, where I was accused of breaking Rule 201 (v), which deals with 'deliberately misleading or endeavouring to mislead racing officials'. What they were *now* trying to imply was that, if my collarbone was indeed cracked as the X-ray showed, I shouldn't have been passed fit to ride in the first place, and that I had somehow misled the doctor on course at Sedgefield.

Barney Curley was so furious by the accusations directed at me that he employed Alan Walls, a top London lawyer, to represent me. Mr Walls tried to plead my case, explaining that leading into and on the day of the race my collarbone had appeared fully healed. It was only during the course of the race, when the horse had started pulling hard, that the injury had been aggravated, causing it to flare up again. But our argument fell on deaf ears. They had already made up their minds.

At the same time, in a separate incident, the hard-talking, high-rolling Barney Curley lost his licence for two years under the notoriously broad Rule 220 (iii) for allegedly 'causing serious damage to the interests of British racing' with his characteristically outspoken remarks. Unsurprisingly, after he threatened to take the Jockey Club to the High Court, he was returned his licence a mere 101 days later.

I proved less fortunate. I was still only learning the ropes of a new trade in a new country, and I had neither the resources nor the smarts that Barney did.

My riding licence was taken away for nearly four months, from 11 June until 30 September. The racing season in England would start back up at the beginning of August, and the suspension meant that I would miss the summer season in Ireland and the first two months of racing in England.

The unfairness of the Sedgefield incident shook my faith.

I got on an aeroplane to Los Angeles and within two weeks I had sat the exam to secure a place at UCLA, while riding trackwork for Charlie Whittingham to help pay for my education. 'You are too good a talent not to ride horses,' the great man had said to me. But in my mind, it was over. I had never wanted to be a jockey in the first place, and the unwelcome environment the stewards had created for me, wittingly or unwittingly, was enough to rattle anyone's resolve. I had tried to soldier on, but I had finally had enough.

It seemed the perfect opportunity to make my move towards a different future.

And then Barney Curley called me, fresh from winning the hearing on his racing ban. 'Come back to England when your suspension is over,' he said to me, 'and I will provide you with a winner.'

I was conflicted. I was at the intersection of two paths and while my head was urging me down one road, my heart was tugging me down the other. This was my moment to break free from expectation and pursue my dreams for myself, and yet I felt an immense sense of loyalty to Barney Curley. Because this is the thing about me – no matter how bad things are, I never let people down. From the start, the bedrock of the relationship between Barney and me had always been mutual respect for one another's abilities and minds. And I couldn't allow myself to betray that trust. After all, this was a man who had publicly said about me, 'If he ever decided to stop riding, I would seriously consider stopping training. There might be as good a jockey around, but none better.'

And despite such high praise, I knew, as did those closest to me, that Barney's admiration for my abilities came second to a subtler, more powerful emotion: responsibility. He shouldered the responsibility for bringing me over from Ireland and his affection towards me extended far deeper than it would have in a strictly mentor–protégé capacity. I had touched some part of Barney's impenetrable heart and he treated me like a second son.

So I agreed to return, only because Barney Curley had asked me to. But I didn't intend to stay for long. In my head, I believed riding Barney's horse would fulfil my obligation to him and then I would return to California, a free man, to live my life.

In the summer of 1989 I went back to England and rode Barney Curley's horse. As he had predicted, the horse was a winner. He was called Above All Hope.

It was a name of great significance because shortly thereafter, for me, all hope did come alive.

My return to England proved to be a turning point in my career. I had decided, almost with certainty, that I wasn't going to stay for long. My plan was to return to America to continue the education that I had always longed to pursue. And so I began to care less about what others thought of me; I began to care more about what I thought of myself. Over the next period, with the solid support of trainers David Ringer, Geoff Hubbard, Neville Callaghan, Philip Mitchell, Mark Tompkins and, of course, the continuing support of Barney Curley, I began to define and own a riding style that was quintessentially mine.

A typical 'Declan Murphy ride' involved settling my mount to travel economically in his cruising speed, getting him jumping comfortably and in rhythm with his stride pattern, and then gradually, almost stealthily, advancing him towards the finish line before finally releasing the power I had reserved, to pounce. This was the point at which Barney Curley would say about me, 'He could drop a horse in twenty lengths further than you'd think he should do, but somehow he would always arrive there to win.'

I was never intimidated or fazed by a situation. It didn't matter how big a race it was, I didn't succumb to pressure – I didn't even feel it. My self-confidence was beginning to surface. This played into my hands and I began to win. My first big victory came in the form of the Irish Champion Hurdle in 1989. I rode Kingsmill, my first winner for the illustrious Tommy Stack, forever part of the Red Rum legend. Later on, in 1993, I would go on to win the Irish

Champion Hurdle for the second time and repay Neville Callaghan for his loyalty in those early years, by winning on Royal Derbi.

After the 1989 Irish Champion Hurdle, things began to turn upwards. There was a casualness in my approach that seemed to disarm people. The more I began to win, the more I began to get noticed. The more I began to get noticed, the more the stewards began to let up. Then, in what seemed a watershed moment in my relationship with the authorities, Bruce Hobbs, senior steward of the Jockey Club (and former Grand National winner), waved a white flag. At his house, over dinner – well behind enemy lines – he said to me, 'You are too good to give up on this. They are trying to get at Barney through you. It will pass, stick with it and your talent will shine through bright. You will rise to the top.'

It seemed, I was good enough.

Bruce Hobbs's words were oddly clairvoyant. Very few get to the very top.

I consider myself extremely fortunate that over the next few years, I was able to carve a place for myself in this elite category. Routinely referred to as having 'the best pair of hands in the business', I rode with startlingly accurate judgement of pace, fuelled by a combination of instinct, attitude and ability. I began to be recognized, lauded and coveted for my seemingly effortless riding style – somehow I made it look easy, and yet I was able to consistently deliver winners.

Race-riding is an art form; a dangerous, demanding art form. To do it professionally and to do it well requires incredible skill. It needs courage and control, intelligence, endurance, strength, perseverance. It needs that intangible *something* – mysterious and marvellous, impossible to define. It takes years to perfect and yet, somehow for me, it seemed to 'just happen', so much so that my brother Eamon once said, 'I was very keen, but Declan was very good.'

There was a thrill in winning, in being *that* good at something. It was addictive. But be that as it may, I could never see myself doing it for ever. I had other plans for life. Life had other plans for me. Racing just kept getting in the way.

And then, once again, fate played its extraordinary hand. I met two people who would forever rewrite the pages of my future, professionally and personally.

Their names were Josh Gifford and Joanna Park.

Each, in their own way, would take my life in a different direction.

A new direction – not right, not left.

But straight up.

And, with that, would emerge a new dawn.

My meteoric rise would begin.

It seemed, I would be good enough.

Four Years

I remember one of the first conversations I have with Ami. It is the moment I tell her that I cannot remember four years of my life. It so happens that I choose to let her in on this little detail *after* we have agreed to write the book. I haven't consciously not told her earlier. It just sort of evolves when she asks about the photos on my walls; she wants to know the story behind each one. I tell her that the photographs are of great significance to me because essentially they are my memory. She looks bemused, asks what I mean. So I tell her. Our exchange goes something like this:

Her: 'What? Gone? All gone? All of it?'
Me: 'Gone.'
Her: 'Who else knows about this?'
Me: 'No one.'
Her: 'Family? Friends? Racing people?'
Me: 'No one.'
Her: 'And now, you're prepared to tell *every*one?'
Me: 'Yes.'
Her: 'Which four years?'
Me: 'The four years right before my accident.'

Her: 'When you were at your most successful – at the zenith of your career? Oh, *perfect.*'
Me: (*laughing*)
Her: 'Trust me, this is many things. But funny is definitely not what I'm feeling right now.'
Me: (*laughing harder*)
Her: 'You want *me* to tell your story when *you* can't tell me your story?'
Me: 'I'll help you. We'll do it together. Don't worry.'
Her: 'Don't *worry*?'

She invents a new word while writing the book – 'Declanism', she calls it; the idealism that is Declan.

The Rise

We all have our secrets.

And I circle back, once again, to mine: I never wanted to be a jockey.

My career was incidental.

But the mere fact that I was a jockey meant I was going to be the best jockey I could possibly be.

It doesn't matter what I do in my life.

It doesn't matter that it may not be what I *want* to do.

If I do it, I will do it to the very best of my ability.

That is the essence of who I am.

So I carried on the charade.

In the summer of 1989, when I returned from California to ride Barney Curley's Above All Hope, I had almost fully decided to give up riding and go back to America to continue my education.

I felt I had repaid my debt of gratitude to Barney Curley. I had agreed to leave Ireland with him when nobody thought I should, I had satisfied my own curiosity of him, I had stuck with him through thick and thin, and I had genuinely learnt more from him than I would have with anyone else. So, I was ready at this point to spread my wings – and go.

Exactly around this time, my presence piqued the interest of one of racing's most revered professionals, the fabled Josh Gifford.

I was still riding Barney Curley's horses at this time, but I had also been riding for several other trainers – Tommy Stack, David Ringer, Geoff Hubbard, Neville Callaghan, Philip Mitchell, Mark Tompkins, to name a few. Until this point, all of them had reached out to me through Barney, either out of courtesy or because there was no other way to get hold of me – I didn't have my number in the book.

In the winter of 1990, Mr Josh Gifford called me direct. He phoned me up, out of the blue, and said, 'Would you ride all my horses at Ascot this Saturday?'

'Yes,' I said, because of who he was. And I did.

After that, over the next several weeks, I started riding most of the Josh Gifford horses. Shortly thereafter, Josh Gifford's stable jockey Richard Rowe retired and, amidst a sea of press speculation, I got offered what many would consider to be one of the best jobs in the country – the opportunity to be his new stable jockey.

Do I go for it?

Behind every story lies another story, and so it is with this one:

It was the year 1983, I was seventeen years old.

Francis O'Callaghan ran the quarantine in Ireland for shipping horses abroad. He was from my village of Hospital so our families knew each other well and I would go ride for him every now and then, to help exercise the horses before they were sent abroad.

If anyone in the racing world truly understood me, it was Francis. He knew I didn't want to be a jockey, but he also knew that I loved horses, I was good with them, they were good with me. While many others struggled with the confusion this seemed to create, he embraced it. He could tell there was something different about me, about my ability to ride, and his support and encouragement for me was unwavering, even at that young age. Francis O'Callaghan went on to follow my career with great pride and I will always look upon him as my first mentor.

One time, Francis took me up to Kilmallock cinema, to watch the movie *Champions*. In the movie, it is Josh Gifford who trains Aldaniti, the horse that the courageous Bob Champion rides to win the 1981 Grand National after he recovers from his illness. I watched the movie, a starstruck teenager. It was my first cinematic experience, and the whole thing affected me greatly – the dark, quiet room, the plush chairs that I could sink into, the deep-red velvet curtains opening to reveal the huge screen, the lifelike images, the colours, the clarity, the music. I could have lost myself in there. And then, of course, there was the story itself. I was so touched, so inspired by Bob Champion's story, by the courage and the bravery of both man and horse. And Josh Gifford came out a total hero – I had so much respect for the trainer for believing in and sticking by his jockey when he could have easily found someone else. Everything moved me deeply, the experience, the story, even the shots of the Sussex Downs where Josh

Gifford kept his horses left me breathless – how glorious England looked!

Now, this same man – this man of the silver screen – had offered *me* the job of stable jockey at his powerful Findon yard. He looked even more of a character in real life than Edward Woodward had portrayed him in the movie. The hat, the bushy eyebrows, he was larger than life, a caricature of himself, a brilliant man of deep convictions, a legendary trainer.

Any way you chose to look at it, for a jockey, riding for a trainer of Josh Gifford's calibre was a huge privilege.

So, do I go for it?

To the outside world, I was Declan: the realist, the pragmatist, the perfectionist . . .

On the inside, I am Declan: the dreamer, the romantic, the idealist . . .

I go for it.

And from here, my ascendancy began.

It was a marked phase in my career – it was The Rise.

Race-riding is all about speed. The ability to achieve speed on horseback. And the only way to internalize this speed, to make it appear natural, is by 'feel'. Therein lies the difference between a good jockey and a great jockey: that innate capability to 'feel' the horse he is riding. So many of us are corrupted by our conscious state but when you block out the peripheral and you focus more on the 'feel', the complete awareness of how the horse moves under you, you transcend this imaginary line – you go from good to great. It really is that simple. As soon as you mount a horse, just cantering a couple of strides, you can learn so much about the animal. Every twitch, every movement means something. And then when you quicken, when you find a horse's correct cruising speed as you start racing, you feel it too – the rhythm of the stride pattern, the beat of the gallop. If you hear it, if you tune yourself to the same frequency, if your horse wins with something to spare, you've got it, it's yours.

At this point in my life, I had it, it was mine. The synergy between me and the horses I was riding was unique – horses wanted to jump for me, horses wanted to win for me. I had honed my skill to a point where I had an uncanny ability to judge the pace of a race, to own my speed, and somehow I found myself in the right place at the right time in every race. People began to become aware of me:

John Cobb said I was a 'phenomenal riding talent'.

Geoff Lester called me 'tremendously gifted'.

John Carter labelled me 'the supreme artist in the saddle'.

Josh Gifford named me 'the most complete rider in the game'.

Suddenly, everybody seemed to want me. It was odd, it was mad. It was just strange.

But despite my perceived success, I remained picky about my rides.

The general industry-wide expectation is that jockeys ride to claim championship honours. This necessitates riding every chance you get. It's a numbers game – the more you ride, the more you can win. My methods were unconventional and sometimes controversial. I wasn't insatiably hungry for the title.

Instead, I preferred to ride the big races and usually only

when the circumstances fit my personal desires. For example, I would have literally paid to ride at the Cheltenham Festival – the right atmosphere always gave me a real buzz – while I preferred not to ride at meetings that felt like they just made up the numbers. So, as unorthodox as I might have appeared to be, ultimately I was who I was – a businessman riding horses.

And so, I never rang up for rides and I didn't have my phone number in the directory. I believed that if someone wanted to book me, they would find a way to reach me. I have always been, by nature, an extremely private person, and although I didn't consciously plan this, it led somehow to an air of mystery about me – no one was really ever able to get their finger on my pulse. Even in my best years as a jockey, and despite my showmanship on the horse, I shied away from the public eye, happily electing for the torch to be pointed away from me as a person, and instead shine on my skill. But even then, I felt no pressure to ride a horse just because the owner or trainer fancied my riding it. I believed that it did no good to ride bad horses for the sake of rides. I rode every horse with 100 per cent dedication, but if I had started riding bad horses for people who didn't know bad horses, I'd have lost that dedication and probably would have been far less effective as a result.

On the other hand, it was always gratifying to ride good horses, and I enjoyed racing through riding for the right people. If I had ever stopped riding for these people, I'd have stopped riding altogether. For me, enjoyment was always key – it is hard to beat the thrill of jumping at speed on an exceptional horse, and this is what I aimed to find in every race. So, I made my decisions for myself – which races I wanted to ride in, which horses I wanted to ride on – not from obligation or compulsion, but from my own volition. I preferred to ride at my own peril and always for a good reason. Anyone who rides professionally knows that racing is predicated on chance; risk and reward. And I wanted to ride when I was sure the risk was worth the reward. Or, to put it bluntly, I wanted to ride when I believed the horse could win.

Because of how I rode, because I chose to pick both my horses and my races, I minimized the degree of risk. I was fairly good at anticipating what the other horses and riders would do, and as long as I had the confidence in my own horse, I felt in control. Everything I did was calculated with one ultimate goal – to make sure the horse and myself did not part company. The result was that I had far fewer falls than my peers. Take, for example, the 1993/94 season, my last before the accident. During this season, the average jockey fell once in every nine rides (as of 2016, the statistic has improved to one in every 14.5 rides) – I had fallen ten times in my 385 rides that season.

In a similar vein, somehow I inadvertently redefined the traditional jockey–trainer relationship. As I advanced in my career, trainers didn't feel the need to give me any instructions. Either they trusted that I knew what I was doing or they thought it unnecessary because I was going to go out and ride the way I thought the horse was best ridden anyway. This wasn't because I thought I knew better than them – I certainly didn't. It was only because I felt that if you had a view, you needed to have the conviction to back it. I had seen people become bitter when they left the decision-making to others. In my mind, the recipe for life and happiness is to decide for yourself. Then the responsibility – of both the decision and the outcome – is yours in totality.

So while I was forever a student of my craft, I had the self-confidence to shout out for what instinctively felt right, even if I stood alone. And just like that, when I followed my gut, I found that things started to fall in place for me, the right opportunities began to present themselves, the right people came my way.

Synchronicity surrounded me.

Josh Gifford recognized this straight away. First and foremost, even in asking me to be his first jockey, he turned the system on its head. In Josh Gifford's world, it was an unspoken rule to promote from within, to promote your own. When Richard Rowe, Josh Gifford's stable jockey, announced his retirement, it seemed a racing certainty that someone already associated with the Gifford stable would take his place. Waiting in the wings were Peter Hobbs, Eamon McKinley, Tom Grantham and my brother Eamon Murphy, who had been joint-champion conditional rider for Josh. Josh Gifford was a traditional man – old school in the way he trained, old school in the way he thought – and although rumours were rife at racecourses up and down the country that he was going to offer me the much-coveted job, it seemed an unlikely departure from convention for a man like him. It was. And yet, he did.

When I was offered the job, as much as I knew it was one of the top three jobs in the country, I accepted it on the rather audacious condition that I would ride Josh's horses directly in the various races; I wouldn't be coming to the yard to ride out on a regular basis, and I certainly wouldn't be moving to Findon from my home in Newmarket.

Josh Gifford thought about this and then he said to me, 'That should work. When Fred Winter rode for Ryan Price, he didn't ride out at the stables either. If it worked for them, it should work for us.'

Given that Fred Winter and Ryan Price were probably one of the most formidable combinations of success in racing in the post-war era, it was clear, just in making the comparison, that Josh Gifford respected my decisions as much as I respected his.

In another exception to the rule, we also agreed on a retainer per horse, as opposed to the standard umbrella contract that would oblige me to ride all Josh's horses, as it was likely there would be horses I wouldn't want to ride. I don't think any other jockey at the time had this sort of working arrangement, but once again, in another notable departure from convention, Josh agreed and, satisfied that it would suit me, I took the job.

Almost immediately, I started riding some very good horses for Josh. Deep Sensation and Bradbury Star, two horses that would be inextricably linked to my best years as a jockey, had previously been good hurdlers for Josh, and were just about to start their first season novice chasing. Both had become high-profile horses in their own right and continued to attract the attention of the race-going public.

I had ridden Bradbury Star to win the 2-mile, 4-furlong Scilly Isles Novices' Chase at Sandown Park in February 1992, and he and Deep Sensation were both going to Cheltenham that March. Josh Gifford had initially wanted to run Bradbury Star in the Arkle Chase, which was being run over 2 miles, but we subsequently decided that the trip might prove too short for him; in all likelihood, he would want further. His only other option was the Sun Alliance Chase, the championship race for novices, which was being run over 3 miles, 1 furlong, and Josh decided to run him in this race instead. I turned up at Cheltenham that year with a good book of rides, but I can unequivocally say that Bradbury Star was the horse that I was most looking forward to riding.

When we jumped off in the Sun Alliance Chase, I slotted into a good position. It was an extremely competitive line-up, but I had total belief in Bradbury Star's potential to rise to the occasion. I thought that if he was able to stay the distance, he would have a great chance at winning the race. In all honesty, however, I had my colours pinned to hope. I willed myself to believe that he *might* stay the trip; I was far from convinced that he *would*. But I felt it was worth a try. As the race progressed, there was a relentless pace set by Mutare and Richard Dunwoody, and sustained by Run For Free and Jimmy Frost. I feared that this might possibly exhaust every ounce of stamina that Bradbury Star had in him. Nevertheless, he jumped impeccably and I delivered him to win halfway up the run-in.

And then it happened, just like I had hoped it wouldn't but feared it might. The elastic band of stamina that I had stretched to its limit, snapped. Bradbury Star ran *nearly* all the way to the line; he was like a car running out of fuel in the last 100 yards. We got beaten by half a length by Miinnehoma, ridden brilliantly by Peter Scudamore.

If anyone needed reminding of how good a jockey Peter Scudamore was, in my opinion, there is no better race than the 1992 Sun Alliance Chase to demonstrate this. He had committed on Miinnehoma a long way from home; such was his skill, he had Miinnehoma running all the way to the line. As for me, I was exhausted when I came back in after the race, more emotionally so than anything, because I had truly believed that Bradbury Star might win. It had been a huge challenge for me, and I'd go so far as to say that it was perhaps the best ride I'd ever given a horse at that point in my career. So the enormity of that sense of expectation and, equally, the disappointment when you get so close to it and are beaten, was huge.

Nevertheless, as I look back on it now, I realize what a momentous event it had been; arguably one of the greatest Sun Alliance Chases that has ever been run at Cheltenham – a window into the future. Miinnehoma, who won the race, went on to win the Grand National two years later. The horse that finished third, Run For Free, won the Scottish Grand National a year later. The horse that finished fourth, Rough Quest, went on to win the Grand National four years later. The horse that finished fifth, Captain Dibble, won the Scottish Grand National in his next run. Bradbury Star would go on to be a two-time Mackeson Gold Cup winner. Rarely in the records of racing history will you find one race where the first five finishers have come out as champions in their own right. So even though we were defeated, I knew Bradbury Star had done his very best and would go on to achieve much, much more.

It was towards the end of my first year with Josh that I decided I needed someone to manage my rides. I had never had an agent before, but with over a hundred horses in the

Gifford yard, my life was becoming complicated.

I didn't feel the need for an agent in the traditional definition of the role; in other words, I didn't need someone to book rides for me. But with Josh having so many runners at so many different meetings, it became clear that I needed help with managing the logistics of my everyday life. Ideally, I was looking for a person who understood the way I liked to work, so they could take over the behind-the-scenes organization and free up my time to ride.

So in my characteristic way, I didn't pick an agent to be my agent. Instead, I picked an agent who wasn't an agent. I chose Marten Julian, a renowned journalist with deep horse-racing expertise. I had been a big fan of Marten for some time and regularly read his *Dark Horses* annuals. I loved the way he wrote about horses; his work showed that he truly understood them – the animal itself, including but not limited to just its role in racing. This appealed to me.

So, I asked him if he would be my manager, and he agreed. Marten had always had a soft spot for me, and I knew he respected my ability as a jockey. I felt instinctively that we would make a good partnership.

But still, as in every new relationship, it took time and experience to learn each other's ways.

This is no better demonstrated than by a humorous anecdote from Marten himself, when Ami asked him to recollect any memorable stories about me during the time we worked together. I may as well reproduce the actual email, from my former agent to my current writer.

This was only a small incident, but it was a big step forward in our relationship. Up until this point, Marten hadn't known me long enough to fully grasp how my mind

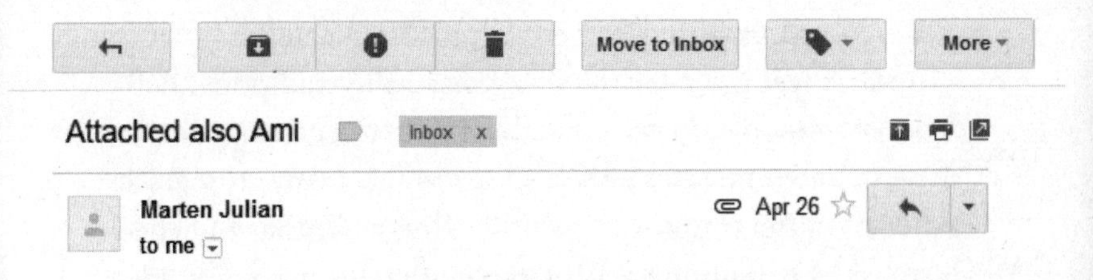

Hi Ami,
I hope you are keeping well.
It was a Saturday afternoon, in the autumn of 1992 when Sedgefield clashed with Newbury. It was very early in my relationship with Declan – possibly a week or two after he appointed me as his agent. I was so keen to impress him and spent hours on the telephone to trainers keen to offer them his services, but I was not aware at that stage of his loyalty to Josh and especially, on this occasion, the owners of the horse. I had wanted to surprise him so waited until I had a full book of rides to tell him that he had five mounts at Sedgefield, all fancied by their trainers. I remember the silence from his end of the 'phone before he said "I have to go to Newbury. The owners of that mare have been very supportive and I can't let them down." My immediate feeling was that he didn't fancy the long trip from his home to Sedgefield.
Nevertheless, I went to Newbury and saw him finish down the field on Josh's mare, popping in and out of the racecourse betting shop to hear (no pictures in those days) three of the intended mounts at Sedgefield win, and the other two finish close seconds.
We were quiet in the car – I wasn't going to say anything – until I recall us passing a fuel station on the M4 when he at last asked how they had run. I told him three had won and the other two would have done if he had ridden them. He smiled and said well done.
That is how I remember it, Ami.
Bye for now

Marten

worked, but after this, he quickly figured me out.

There is a truth about me: I regret losing more than I enjoy winning.

Marten understood this. And he understood that because of this, I was very particular about picking horses that I believed could win for me. But he also understood that it wasn't always quite so straightforward. Loyalty to those who mattered to me was sacred – above all else, and at all costs. And oftentimes with me, 'predictable' broke down.

In the case above, I had made my decision against Marten's better judgement, driven by an intangible force greater than my love of winning. And when, as it so transpired, the outcome didn't go my way, I was happy to accept my loss graciously, without any regret or bitterness. My wins were my own and so were my losses. Whether this was a character trait or an idiosyncrasy, I don't know, but Marten recognized it and accepted me for it.

And so it didn't bother me in the least that Marten had never been an agent. It mattered to me that we got along well and that he knew what made me tick. I had an understanding with him; a language we spoke when we were together that was special. We could sit down, Marten and I, and talk about horses all day long. If you are lucky, you meet people in your life that bring out the best in you, that really get you to perform beyond yourself. Marten was one of those people to me; a firm and loyal friend.

It is only fitting to mention at this point that the job I had with Josh Gifford could not have been as seamless as it was, had it not been for two men (apart from Marten) – my brother Eamon and Richard Rowe, Josh Gifford's stable jockey before me. The advice Richard Rowe was able to give me when I was riding those horses I'd never ridden before, which he'd been riding, was invaluable. As for Eamon, he also rode for the stable but the difference between us was that he spent a lot of time there and I spent no time there. I would harass him endlessly for his advice on the horses and he would dispense it liberally, openly and willingly; it was above and beyond any call of duty, and I consider myself fortunate to have had such a great friend in such a great brother.

It was during the Josh Gifford years that my career made the quantum leap forward. I had perfected a riding style that mirrored my personality – stylish, in control, and with complete self-belief. These remained my defining traits not only in the way I raced, but also in my way of being.

In fact, it was this same self-confidence that ultimately lay at the heart of the other significant event in my life at this time. Josh Gifford was but one of the driving forces behind my decision to stay on in England.

The other was destiny.

And what divine destiny it was. I was overcome by the heady forces of love, that most powerful emotion of them all.

I met Joanna; I was intoxicated.

And I found myself stuck at the crossroads with The Golden State on one side and a beautiful girl on the other. Naturally, I chose the girl.

A Beautiful Girl

The first time I saw her, Joanna was riding a horse at Charlie King's livery yard in Newmarket. As I rode past, I couldn't take my eyes off her; I had never seen someone so beautiful sitting on a horse. In my mind, only people who looked like me rode horses, and I remember thinking, This doesn't look right, she's too beautiful to belong on a horse.

Her eyes looked up then, met mine. She almost smiled, and I almost fell off my horse.

In My Wake

People say death is much harder for those left behind than for the person who dies.

I am fully qualified to corroborate this hypothesis.

While the events detailed below were unfolding across Ireland on my account, I lay oblivious to it all, in a private world of my own.

A world without fear or pain.

Within an hour of my accident, my siblings who lived in Ireland – my sisters Kathleen and Geraldine, and my brother Laurence – had rushed home to Mam and Dad.

Five anxious people now paced up and down the sitting room of 5 Bank Place, with no inkling of what fate awaited me. Turn by turn they worked the phone, desperately trying to call anyone they could think of for answers, but found no one who could give them any.

According to Kathleen, in the immediate wake of my fall, my parents became silent and withdrawn, revealing little emotion as they retreated into their individual, impenetrable worlds of self-preservation.

Mam busied herself doing everything she could around

the house, trying to keep her mind distracted, but my sisters, who knew her so well, could tell she was concealing some deep, heavy sadness that she didn't know how to confront.

Dad tried to use logic to cope. For him, every attempt at rationalizing what had happened required the need to impart reason into something that would otherwise be too amorphous to handle.

So he sat in his chair and announced my invincibility to no one in particular. 'He'll be fine, he'll be fine, he'll be fine. He's young, he's strong, he's fit, he'll be fine,' he would proclaim, as if, by repeating the same words over and over again, he was willing them to be true.

The games we play . . .

The first update came at 11.30 p.m., close to nine hours after my accident.

When the phone rang, my sister Geraldine answered it. It was my brother Pat, calling from the hospital in Liverpool.

When he spoke to Geraldine, she could hear the raw emotion in his voice, even as he tried to disguise it. But he didn't mince his words: 'I've just been in to see him. It's not looking good. I'm not holding out much hope. Don't tell Mam. Let me speak to her.'

As soon as Geraldine passed the phone to my mother, Pat's voice on the other end of the line took on a calm, almost carefree tone. 'Mammy,' he said, 'he's going to be fine. Don't listen to what anybody else says. Don't read the papers and don't listen to what they say on the news.'

'Oh, Pat, is he going to live?' she asked one son, about another.

Pat looked up then at the large, round clock on the hospital wall. It was nearing midnight and he knew his parents would get scant sleep even under the best of circumstances. So he crossed his index and middle fingers down by his leg – a childhood habit – as he lied, 'He'll be fine, Mammy. I'm telling you he'll be OK. He's bad, but he'll be fine.'

And for the first time that day, my mother opened her heart and wept.

Following Pat's call, Geraldine, Kathleen and Laurence began to discuss the logistics of getting Mam and Dad to Liverpool.

This might have seemed straightforward; it was anything but.

My parents were torn in two.

While a part of them felt they had to be at the hospital, another part was too afraid of what unknown beast they were choosing to confront. 'I'm not sure I can face it,' Mam kept saying, almost to herself.

My sisters were unanimously against Mam and Dad making the trip across to England, praying silently that they wouldn't insist. Their motivation was simple: they wanted to protect our parents from as much as they could for as long as they could. Both girls had sons of their own; they understood, like any parent, the monumental burden of being one. There is never a way to temper the trauma of seeing one's own child at death's door. Hearing about it is one thing; seeing it in the cold light of day would be another.

And so at four in the morning it was decided, if not by

consensus then by some resigned understanding: Laurence and Geraldine were to go and Kathleen was to stay in Ireland with Mam and Dad.

For the next ninety-six hours, friends and strangers across Ireland rallied around my parents in an unprecedented show of solidarity. The response was overwhelming. People were coming in and out of the house constantly to voice their support for my parents, many times, if only to say that I was in their prayers.

Five Bank Place became redolent with the scent of flowers. Cards and letters poured in. Bells tolled as Masses in my name were held in churches across the country.

While 296 miles away, I lay blissfully ignorant in a deep, dark sleep.

Stranger things have been known to happen, but in a small Irish town in Limerick, a sixty-four-year-old Catholic nun was trying to find the God that had found her.

Sister Bridget from the Presentation Convent had taught me at school when I was barely four or five years old, and watched me grow from a child to a man with both great interest and great fondness. Over the years, we had become firm friends and allies. Right from my pony-racing days and into my years as a professional jockey, she followed my career intently, cheering on my successes with heartfelt enthusiasm.

At 4 p.m. on Monday, 2 May, Sister Bridget heard the news of my fall. At 7 a.m. the following morning, she came to the house to pray with my mother. And then she came at that same hour, every single morning, for weeks. She did this to share her faith, to bring peace to my mother's torment, and more than anything else, to make sure everyone believed that, in this hour of need, *her* God was on *my* side. Ave Maria.

On the morning of 3 May, when the newspaper came in, my mother refused to read it. 'Pat said no,' she told my father, shaking her head. 'Pat said no, so I'm not going to.'

It was just as well. According to the papers, I had died before I had even left the racecourse.

But it didn't stop my father.

We had an open fire in the kitchen, which he lit himself every morning and every evening religiously, except in the summer months. It was in front of the fire, with a cup of tea and two biscuits – never more, never less – that he always read the morning paper.

He did the same on the day in question. Even as the blood drained from his face, he didn't stop until he had read it all, every last gruesome detail . . .

Then he tore it up and fed it to the flames.

It was the same with the television.

Sky News was relentless. My fall off Arcot was replayed as 'Breaking News' every hour for days on a never-ending loop, while the words 'Irish jump jockey fights for his life' flashed impatiently across the TV screen. Carefully chosen words, words that said everything, while still saying nothing.

The first time she saw it, Mam looked at Dad with fierce resolve in her eyes.

'If he makes it, he's never again to sit on a horse.'

After that, no one could bear to have the TV on.

The phone became an object of fear.

Mam became terrorized by it, recoiling every time it rang, refusing to answer it, saying to Dad, 'You get that, I can't.'

It became so frightful that she would walk around the sofa in the living room just to avoid walking by the phone, afraid that it was going to ring and say the wrong thing.

On Thursday, 5 May, a neighbour called on Mam and Dad to offer his condolences.

He had been told in the village that I had died and had come right away to make sure that my parents were OK.

'I'm very sorry to hear about Declan,' he said as he walked through the door.

Willie Scan Ryan had lived next door to us ever since I could remember, and he and my parents had been very good friends for years and years. In fact, it would be fair to say that he would have been as proud of me as Mam and Dad were, routinely phoning me up to congratulate me after a win, or simply to have a chat.

So it could be stated with absolute certainty that there would have been no trace of malice in his words, only genuine sadness. But it was enough to break Mam.

Our mother was one of the softest, most gentle of people, never one to lose her calm. No one – not even her own children – had ever known Mam to get cross.

That day she lifted her chin bravely, looked the neighbour in the eye and said with a steely anger, 'Get out. Get out. Enough. He's not dead. Get out.'

*

At lunchtime on Friday, 6 May, my brother Laurence rang and asked for my sister Kathleen.

He was brief. 'It's going to be a miracle if he survives. Don't tell Mam and Dad.'

Kathleen didn't, mostly because she didn't know how to. Someone else, however, did.

Within an hour of Laurence's call, the doorbell rang. When my father answered the door, he was greeted with words he never thought he would hear in his lifetime. Standing at his doorstep were two men, who introduced themselves as a reporter and a photographer.

'Good evening, Mr Murphy. We've come to get your reaction to Declan's death,' the reporter said, while the photographer clicked photo after photo of my father's horrified face.

It was a few minutes before my father gathered his wits about him and slammed the door in their faces.

On the same day, a few hours later, a strange solitude descended over 5 Bank Place. It was that rare window when there were no visitors. Kathleen had gone back home to her young boys for a few hours. Tommy and Maura were alone.

Tommy was trying to read the *Irish Independent* in a vain attempt to get his mind off the darkness that had enveloped their lives.

When the phone rang, Maura was walking right by it.

She flinched. She had been avoiding the telephone for days, but now she was trapped in a bizarre conundrum of proximity. So she looked over at my father, then, hesitating only slightly, picked up the receiver with a small defiant nod.

'Hello?' she said. Followed by, 'Oh, hello, Pat.'

Tommy sat upright, leaning in to try to hear the conversation, to read Maura's face. But her face revealed nothing.

When at last she opened her mouth, it was simply to say, 'Thank you, Pat.' Then she took the receiver away from her ear.

Tommy could still hear Pat's voice talking on the other end of the phone line, but Maura put the receiver back on the phone in a slow, deliberate move. Then she picked it up again and left it off the hook.

Still looking down, she spoke, her voice unnaturally steady. 'They want to turn the machines off.'

Tommy felt his body crumble beneath him as he slumped into his chair. He felt a sudden wave of nausea sweep over him, followed by pure terror. He shut his eyes. When he opened them, Maura was still standing in the same place, eyes down, fixed on the phone.

When she finally looked up to meet his gaze, there was a faraway look in her eyes, a calm that took Tommy's breath away.

A ray of sunlight illuminated her short fair curls, framing her face like a halo.

Tommy Murphy had never seen his wife look like that before – how vulnerable she appeared. Angelic almost.

Slowly, he stood up and walked towards her, his face betraying none of the pain that was tearing through his body. He put his arms around her, not so much because he thought she needed it, but because he did. They stood there for a long time, just holding each other.

Two Picassos

We sit down, Ami and I, to 'do the book', as she calls our book meetings. She's uncharacteristically quiet.

'What's wrong?' I ask.

'I want to do the races today,' she says.

'So let's do them,' I offer helpfully.

'I don't know how to write up races, Dec . . .' she says. 'And you can't remember them.'

We are both silent for a while. Somehow, in this moment, silence seems most appropriate.

Eventually, I am the one to break it. She won't, so I must.

'Right,' I say cheerily, 'how about we get all those news clippings and go through them together?'

She sits cross-legged on the floor of her living room. Spread out in front of her is a mountain of newspaper cuttings, arranged in chronological order, grouped by race. On her computer are the YouTube videos of almost every race I have ridden in, all queued up in sequence, from earliest to most recent, ready to play when we are. She is undaunted by the amount of work we have ahead of us – no stone must be left unturned.

Hundreds of hours later, when we've been through the

lot, she says, 'I don't want to feature a race for the sake of featuring a race. I want a race with a story.'

'A race is a race. How can a race have a story?' I ask, genuinely perplexed.

'It must,' she says, picking up a cushion from the sofa. 'Even this cushion has a story. Everything has a story. Think.'

And so I do.

I think.

And I pick out for her a few of my races 'with a story' – the Tripleprint Gold Cup, the Queen Mother Champion Chase, others that have 'meaning' in the way that she expects. Then we reconstruct them like two Picassos, frame by frame, fence by fence. We objectify, analyse and fracture. We disrupt the notion of traditional autobiography, we mock the 'I', we take apart the neatly packaged illusion of perspective. And then, once we have adequately destroyed, we proceed to rebuild. We begin to create *Centaur*. We connect dots, draw lines and take leaps. We reconstruct and reassemble, fusing together past and present, my life, my truth, her perspective, her biases, slowly stitching together the patches of pride that made up those years of glory. Little bits come back to me as I relive the races, but whether it is memory or imagination, I will never know.

Weeks later, when we are finally done, she reads them out loud to me. 'Be brutal,' she says. 'Don't worry about my feelings.'

I listen, my eyes closed. She's taken me back. I feel like I'm there, riding.

When she's done reading, she looks at me expectantly.

'I think it's very good,' I say. 'Well done, you should be very happy.'

'*You* need to be happy,' she says.

'I couldn't be happier,' I say.

'That's what counts,' she says.

I pause. And then, 'You know, Ami, I can't remember the feeling of winning a race.'

Zenith

It was the day of the 1993 Tripleprint Gold Cup. I was riding Fragrant Dawn for Martin Pipe, British jump-racing champion trainer, one of the most successful and influential people in the history of the sport – a man who revolutionized the world of racing in many ways.

The race was being run over 2.5 miles, the horse had never run past 2 miles, and nobody thought he would stay the distance. I had never sat on the horse before, but I'd watched him run, I had even beaten him previously in some good races. Instinctively, I believed this was enough to give me a feel for what he was like, what drove him to win.

That afternoon, before the start of the race, Martin Pipe wandered up to me in the paddock and asked me if I thought Fragrant Dawn would stay the distance. I turned to Martin Pipe, the trainer who devoted himself to training this horse – this man who knew the animal's personality, his character, his capability, more intimately than anybody else in the world – and I said to him with quiet confidence, 'If you let me ride him how I think he's best ridden, the trip will not beat him.'

Martin Pipe didn't say anything more. He didn't give me any instructions. He only smiled.

As I was cantering down to the start, I focused my mind on trying to read the horse underneath me – the minutest of movements told a story, and I learnt everything I needed to about Fragrant Dawn just in those few minutes. He was a fervent horse, a veritable bundle of flames, and I thought to myself, I need to contain this horse's energy, get him to travel comfortably within himself and try to keep a lid on his enthusiasm as much as I can, so as to get the best out of him when I most need it.

So I jumped off on Fragrant Dawn, level with all the other runners, but by the time they had cleared the second fence, I was trailing close to a hundred yards behind the rest. People seemed amazed that I was letting the field get that far away from me. As if he was reading my mind, at this precise moment, Sir Peter O'Sullevan, racing commentator for the BBC, announced, as we jumped, 'It's Young Hustler, Second Schedual and General Pershing . . . and all safely over it . . . with Fragrant Dawn just the back marker.'

But that didn't bother me in the least. I knew what I was doing. I knew exactly what I was doing.

The most important thing for a jockey is to never give a start in a race. It doesn't matter if you're going to drop your horse in, you always break level with all the other horses until you've got into a rhythm with your horse's stride pattern, until you've obtained the horse's cruising speed. You might find the other horses getting away from you because the pace of the race is very fast, but the crucial thing is to stay confident in your own pace.

In this case, contrary to outside perception, the inside reality was that I was fully in control of my position. In fact, I wasn't letting the field run away from me, I was merely maintaining a pace that I knew my horse could sustain. And Fragrant Dawn and I had the perfect stride at every fence. Then, just as I had predicted, by the time we had jumped ten fences, halfway down the far side, the horses that mattered started to take position and the dynamic began to shift.

As we approached the top of the hill at Cheltenham, five fences from home, I had crept into fifth position. But I still hadn't changed gear – I was relaxed, in complete cruise control. Then, at that point on the track, just before you start racing downhill towards the third last fence, I found myself getting back into the game. And, as was characteristic of my riding style, I did this with apparent effortlessness. People said I made it look easy. Here's a secret, though – there was none. There was never any deep, dark secret. It looked easy because it was easy. It was easy because I simplified it. The art of my race-riding was the ability to ride the race to suit my horse and not my horse to suit the race. This remained the defining characteristic of my riding style, the one constant that set me apart; my trump card.

In that race like any other, all I was doing was maintaining my horse's cruising speed. And at that pace, I turned into the straight with two fences to jump. There were three horses ahead of me but I looked over my shoulder to see if there was any danger behind me, such was my confidence that I would beat all three in front of me.

And then, for the first time in the race, I changed gear – I quickened half a stride. My move was subtle, but Sir Peter O'Sullevan, brilliant and shrewd as ever, clearly picked up on this, as he announced, 'Coming down to the second last now in the Tripleprint Gold Cup, and as they do so, it's Young Hustler and Second Schedual, but Fragrant Dawn is still breathing down their necks, a very close third.'

And so I jumped the second last, moving (according to the commentary) 'threateningly' into second place, still four lengths down on the leader ridden by Carl Llewellyn. But I was somebody so in control of my environment that I remained unfazed; I knew Fragrant Dawn had more to give. The leader, Young Hustler, was a horse with a big reputation, and yet I had the pluck to sit four lengths off him and take a bold leap at the last. I won by a length and the horse never came off the bridle.

Fragrant Dawn had run a spectacular race. Nobody had thought he could do it. I believed – if ridden the way he should be ridden – that he could and he would. He did. We ghosted to victory.

It was my opinion against that of the rest of the world and at such times there are usually only two outcomes: you either come out a hero or an idiot.

I went out to ride the race on a horse I'd never sat on and I rode a race to remember. Everyone has their moment of glory. That was mine. I was no hero, but at least I wasn't an idiot.

Curiously, in the Queen Mother Champion Chase that same year, with the very opposite set of facts, the result was the very same as in the Tripleprint Gold Cup. Josh Gifford did not want to run Deep Sensation in the Champion Chase because he thought the horse wanted further – 2.5 miles instead of the 2 miles in the race. Very simply, he didn't believe Deep Sensation had what it took to win the race.

I had ridden Deep Sensation at Wetherby, in a trial race – a dress rehearsal for the Queen Mother Champion Chase – just four weeks prior. Circumstances hadn't suited him on the day and he got beaten, finishing second. I had ridden the race so as not to affect his confidence, but I had learnt something from it; our defeat had taught us an invaluable lesson. I realized that Deep Sensation was a complicated ride; as a jockey riding him, you couldn't go and make your own pace because he was a reluctant horse at the best of times. Instead, he needed to be carried into a race by the other horses. At Wetherby, there had been insufficient pace in the race. I knew – with a fair degree of certainty – that this would not be the case in the Champion Chase.

When we returned to the second enclosure that afternoon at Wetherby, I walked up to Robin Eliot, Deep Sensation's owner, and I said simply, 'This horse will win the Queen Mother Champion Chase.'

Robin Eliot looked at me, bemused. Then he chuckled nervously and gave me a pat on the back. I couldn't blame him for his scepticism – it was a bold claim from someone who had just been beaten.

But a good jockey isn't one who never makes mistakes; it's one who has the ability to learn from them. I had learnt enough about Deep Sensation to carry with me that level of certainty. I knew he wanted sun on his back, he wanted good ground and he needed maximum pace. And I knew in the Queen Mother Champion Chase that year there was going to be a lot of pace. I was confident this would play into my hands. I believed in Deep Sensation. I believed that when the conditions were right, this horse was better than the rest of them.

I made my case to both owner, Robin Eliot, and trainer, Josh Gifford. I said, 'In the Queen Mother Champion Chase, there will be a lot of pace because Howard Johnson's Boro Smackeroo will run, and if Boro Smackeroo runs in the Queen Mother Champion Chase, we will win the Queen Mother Champion Chase.'

Robin Eliot was somewhat convinced; Josh Gifford was not. The last thing Josh Gifford said to me – and this will remain my abiding memory of this man, my friend, one of the most respected trainers in the country – as he was legging me up in that race, was, 'We're in the wrong fucking race, but good luck anyway.'

And with those words in my head and a smile on my lips, I cantered off to the start.

To many, the Queen Mother Champion Chase is considered the ultimate test of speed and agility. That year it was being touted as a two-horse contest – Katabatic and Waterloo Boy were the top two contenders, well-backed and strongly fancied to win. When we were down at the start, I was exchanging a laugh with Peter Scudamore, who was joking with me aboard Cyphrate.

He said, 'Declan, you and I are playing for third fiddle at best!'

I laughed with him, but third fiddle couldn't have been further from my mind.

I had different ideas.

We jumped off in the race and as I expected, Boro

Smackeroo initiated a phenomenal pace on fast ground at Cheltenham. It was music to my ears. Below me, Deep Sensation was jumping magnificently. Halfway down the back straight, Boro Smackeroo had a commanding lead, but the beauty of this for me was that the other horses were following him closely, not letting him get away. Perfect, I thought, they are playing right into my hands.

At the top of the hill, Boro Smackeroo was in the lead, Peter Scudamore on Cyphrate was close behind him on the outside, Katabatic was outside him, Fragrant Dawn was in between them, and I was trailing behind, thinking, Boys, you guys just tussle for the lead, I'm happy to stay here and pick it up from whichever one of you eventually carries me to the last.

We ran down the hill and I eased out to the centre of the fences. Cyphrate took over the lead from Boro Smackeroo and Fragrant Dawn followed Cyphrate through. I sat in behind, jumping the second last and went between them turning into the straight.

Truth be told, I was under pressure at this point – Deep Sensation was an intelligent horse, but a reluctant one – he wasn't going forward for me in the way that I wanted. I thought of Josh Gifford's words on the one hand, and my conviction on the other. And I stood my ground. Deep Sensation may have been temperamental, but he was talented. I had already brought him to his cruising speed. I knew he would deliver for me *if* I was successful in bringing out the best in him from this point on. The trick was not to commit him, not to let him think about the race. In fact, my goal was to not even let him know he was square in the middle of one.

Going into the last fence, I got up the inside of Fragrant Dawn and just on the outside of Cyphrate, as I thought this would serve to motivate Deep Sensation by giving him cover. Such was the rhythm I got him into, that Deep Sensation responded in exactly the way I had hoped – he crept forward, his competitive spirit burning, and quickening up the hill won 'the wrong fucking race' by three-quarters of a length.

After the race, Josh Gifford would say about Deep Sensation, 'Declan has made a man of him. He has always been one of my favourites. This is my greatest thrill after Aldaniti.'

Victory is its own reward. But for Josh Gifford to compare my win to Bob Champion's victory in the Grand National was more than I could ask or imagine.

I was king of the castle on St Patrick's Day.

So, when Ami asks me to recount races with 'meaning', these are the ones I pick. And I pick them not because they showcase my horsemanship at its best. I pick them because they signify an innate character trait that defines me. I pick them because they demonstrate, far more accurately than I could put in words, my iron-clad belief in the decisions I make – and then, my will to follow through, no matter the odds.

Because at the end of the day, decision-making is what sets people apart in this profession – believing in something, making a decision and then owning that decision. All the better if you are making a call that nobody agrees with – eyebrows are raised along with the stakes, and you find yourself alone in the middle of a desert, with no cover. In both races, the Tripleprint Gold Cup as well as the Queen Mother Champion Chase, I was staking not only my reputation, but equally the owner's, the trainer's and the horse's. If I had been proven wrong, there would have been no burying my head in the sand, the responsibility for the outcome would have been mine alone. But the water from an oasis always tastes sweet. So, without a doubt, it had been a gamble, but I believed that it was worth it.

I believed in the horse. I believed in my ability as a horseman.

This is what defined me, what lay at my core: I believed.

No matter what, I always believed.

And belief is the supreme emotion.

The bigger the obstacle, the greater my resolve.

The bigger the race, the better I was.

The 1993/94 season was my best yet. In the twelve months before my accident I won more races than I could dream of.

I won the Irish Champion Hurdle, I won the H&T Walker Gold Cup, I won the Queen Mother Champion Chase, I won the Melling Chase, I won the Mackeson Gold Cup, I won the Tripleprint Gold Cup, I won the Bula Hurdle, I won the Lanzarote Hurdle, I won the Cheltenham Silver Trophy Chase.

I was riding on the crest of a mighty wave.

Everything I touched turned to gold.

I did not win the Swinton Handicap Hurdle.

Instead, I lost my career and almost my life.

I was labelled the jockey who cheated death.

'Great,' you say, 'truly great. You cheated death, that's remarkable. But . . . do you feel like life cheated you?'

'No,' I say. 'No chance. Because *I* was cheating life by pretending to be a jockey in the first place.'

She's right, you know, my ghost. Everybody has sins.

A Helmet and a Phone

It is a safe house for these men who choose to live dangerously. The place where they dress to perform; the place where they undress after they perform. A sacred sanctuary of sorts where horsemen bond, the only place on the racecourse where rivalry bows to friendship.

It is a big boys' locker room, the jockeys' weighing room – you walk in and it hits you, the energy and intensity and testosterone; it engulfs you.

It is the smell that gets you first, the overpowering smell of leather and saddles and sweat . . . and men.

Visually, it feels cluttered, but there is complete order to the chaos. Wooden benches run along the perimeter of the room, saddle racks hang neatly in a row on the walls above, clothes hooks over the benches, shoes below. In the middle of the room are large wooden tables where the valets do their work.

The atmosphere is jovial. Alive. There is plenty of back-slapping and hand-shaking and camaraderie and laughter. The noise level grabs you, the snippets of conversation typical of what one would expect of a group of high-achieving guys thrown together in a small space – there's talk of horses and Julia Roberts and trainers and movies and sex and girls and cars and more girls. If someone with authority walked in and demanded silence at the top of his lungs, it would take a good five minutes to achieve his intended result. If he was lucky.

Today, all this was different.

Today, there had been a catastrophe of immeasurable magnitude. One of their own had suffered a fall and nobody knew whether he was to live or die.

There was no buzz, no activity, just the defeated air of unspeakable loss. Moments ago, there had been live coverage from the BBC on the wall-mounted TV screen, showing scenes of the fall, real-time updates on the condition of horse and rider. Then, suddenly, the images stopped, cutting straight to a sober-faced Sir Peter O'Sullevan, who announced gravely, in his distinctive, measured tone, 'We will bring you news of Declan Murphy when we have it.'

A subdued silence descended. What did this mean? Why was there no more news? Could he be . . . ? No, he couldn't possibly. Or could he? No one wanted to think the word, let alone say it out loud.

A few minutes later, in the midst of a hush rarely ever felt in a jockeys' weighing room, the course doctor entered. He had objects in his hand: my colours and helmet, dyed an unnatural shade of red.

Brave men averted their eyes.

The objects placed wordlessly on the table, the unfortunate harbinger of gloom departed swiftly, knowing his place, leaving the jockeys with the privacy to grieve. Blood seeped from the helmet, running a long, thin river across the table, then spilling over the edge in slow, rhythmic drops.

A portent.

Silence screamed with horror. The sour, metallic stench of death spread languidly, as if in no rush to fill this space it now claimed as its own. The walls closed in. Thirty horsemen stood frozen in the moment, as though confronted by something not of this world. The air hung thick and heavy, like a jockey's lead cloth.

Then, the silence was jarred – suddenly, unexpectedly. Absurd tones of a happy melody filled the room.

An uneasy shuffling of feet, the dull murmur of voices; the awkwardness of the timing of it all. Irritation now, annoyance even – whose phone could it be and why was the owner not claiming it?

Heads turned this way and that, and finally converged on one corner of the room, on what was my spot on the bench, where all my things lay, eerily neat, exactly as I would have wanted to find them when I returned from my race.

Jockey Ross Campbell, standing nearest, moved forward, hesitating only slightly before he rummaged through my bag to retrieve the damning instrument.

In his hand, the phone continued its merry jingle.

In his eyes, he might as well have seen the devil himself.

When he opened his mouth to speak, it was barely a whisper. 'It's Joanna,' he said, looking down. Then, shaking his head, he passed the phone to the man next to him. 'I'm not answering it.'

The man next to him recoiled, hands in the air – *not me.*

The third person proved more gallant, thumb moving determinedly towards the green 'talk' button. Then, he weakened, as the full weight of what he was taking on sunk in. 'Fuck, I'm not answering it.'

The cursed, obstinate gall of it all.

And so it was passed from man to man, like a child's game of pass the parcel, played to the tune of the ringing phone.

Brave men lost their mettle.

You answer it. No, you answer it. For fuck's sake, someone answer it.

Then, abruptly, the ringing stopped.

And mercifully, there was silence.

The Lawyer I Never Was

Remember how I said I never wanted to be a jockey?

Remember when I said I always wanted to be a lawyer?

The King George VI Chase on Boxing Day is one of the most prestigious races of the jumping calendar. On 27 December 1993, I was riding Bradbury Star over a trip that I realized was just beyond his best, and to have any chance of winning, I knew I would have to give him a really economical ride and use all the momentum I could find to have him running to the finish.

At the second last we jumped five abreast. Barton Bank and Adrian Maguire were getting into top gear, The Fellow and Adam Kondrat quickened yet again, Young Hustler and Carl Llewellyn were tapped for toe. Going to the last fence, Barton Bank was really galloping, Bradbury Star was still happy enough and The Fellow was cooked.

I concentrated on jumping the last fence well before committing my horse, and hoped he would find enough to get past Barton Bank. He did find enough, but Barton Bank found more.

I did not pick up my stick until I felt it really necessary, which was 100 yards from the line, and used it to encourage my horse to dig a bit deeper. In the end we were beaten by a short head by a courageous horse with a lot of class, who had been given a marvellous ride by Adrian Maguire.

It was considered by many to be one of the best finishes to a big race in years. In Donn McClean's words, 'December 27 saw one of the greatest King Georges in living memory, one that will be remembered for the titanic struggle between two men and two beasts. Four champions.'

But despite it all, the local stewards at Kempton decided that the riding didn't quite make the cut for them. They handed both Adrian Maguire and myself a two-day ban because they considered that we had 'not given our mounts enough time to respond to the stick'. I knew expressly that this wasn't the case. Neither Adrian nor I had abused our position as professional jockeys. We used our whips in rhythm with the horses' strides, gave them time to respond and made contact in the correct place. It wasn't anything other than what could be reasonably expected from a jockey in a race.

The Jockey Club's whip instruction (H9) was topical at the time – there was considerable inconsistency among the various steward panels at different racecourses on what the rules were, and bans were being administered arbitrarily without full clarity on the exact nature of the committing offence. I felt, on principle, that it was about time the issue was resolved once and for all. I knew with certainty that every jockey in the weighing room had worked hard on the new guidelines and rode with them in mind, and I believed the Kempton stewards had misinterpreted both what had happened as well as their own instructions – only a jockey can determine how much time a horse needs to respond and whether it is responding.

So, in a move aimed at forcing a review of the Jockey Club's controversial whip regulations, I decided to appeal against the ban. I wanted to do this not so much to clear my own name, but because I felt that the reputation of jockeys, as a whole, was being tarnished in the public eye.

I wasn't alone in my feelings. The ban imposed on us – and what it stood for – outraged trainers and jockeys across the industry, and there was an outcry against the inconsistencies in the application of the guidelines.

Toby Balding, chairman of the Jump Committee of the National Trainers Federation, accused the Kempton stewards of 'nit-picking'. He said, 'Two highly competent jockeys were riding at the top of their ability. If that was wrong, then we'd all better give up.'

Barton Bank's trainer, the Duke (David Nicholson) said, 'It is a great shame that a marvellous horse race was spoilt by a stupid and scandalous decision from the stewards.'

However, neither he nor Adrian Maguire chose to appeal against the suspension.

Not so me.

In January 1994, I had the opportunity to wear that tailor-made suit and those shiny new shoes, and become someone I had always dreamt I would be – the lawyer I never was. In a bold move that shocked many, I decided to dispense with legal counsel and represent myself in appealing against the ban. Nobody thought I had a chance of winning, but with every shred of visual and professional evidence supporting my case, I was totally confident of the outcome.

At a ninety-minute hearing, I, along with the members of the Disciplinary Committee, viewed head-on camera evidence in slow motion that showed exactly when I made contact with the whip, and that my horse had two or three strides to respond. This, I felt as a professional jockey, was adequate time to know whether he was responding. I argued that I had nothing against the H9 – in fact I thought it was well-intentioned and good for the sport, but only if guidelines were taken as guidelines, and applied with flexibility and consistency. However, if the guideline was to be administered not only as a rule, but as a rule of such intensity, then it highlighted the need for professional stewarding to enforce it.

I spoke; they listened. And with that I overturned a decision, described rather prematurely by the Kempton stewards as a 'cut and dried case'. The three-man Disciplinary Committee upheld my appeal and I walked out of the inquiry having been fully exonerated.

Anthony Mildmay-White, chairman of the Disciplinary Committee, speaking at a specially convened press conference after the hearing, said, 'We had the benefit of ample time to consider the appeal whereas the Kempton stewards were under great pressure with another inquiry stacked up behind them.'

Nothing more needed saying.

The episode attracted significant industry-wide interest. There were those who considered I had done more for the good of racing than I might have realized. I made history by becoming the first jockey to dispense with legal counsel and represent myself, and to this end I received letters from several trainers, jockeys and owners across the country, congratulating me on a 'very brave decision that had resulted in an important victory for racing'.

Little did I know at the time that it would be this conviction of mind, this innate ability to weigh the risks and make non-consensus calls, this confidence to stand alone, even against respected minds with formidable reputations – and be proven right – that would later serve as my greatest strength and save me in my darkest hour.

Howard Wright said about me, 'Anyone who has spent more than five minutes in Murphy's company discussing life in general and his riding career in particular will testify that he knows his own mind. Confident and articulate, he is his own man, nobody's fool. He will not waste a dozen words when half as many will do the job as well. Not more than a handful of jockeys would be as capable of facing the inevitably intimidating atmosphere of an inquiry bereft of assistance, but Murphy is, and did.'

David Pipe, director of public affairs for the Jockey Club, sent me a letter thanking me for my 'constructive and sensible remarks to the press'.

For my part, I took out of that inquiry not only a success but a friend in Anthony Mildmay-White.

My two-day ban was lifted. My conscience clear, I announced to the press that I would be taking the two days off anyway . . .

The Fall

Citius – Altius – Fortius
Faster – Higher – Stronger

The Olympic motto: the essence of an elite athlete.

They are born with this, this inherent *need* for speed, for strength, for fortitude; it is ingrained in their DNA. They spend their lives in the relentless pursuit of ultimate performance. They are consumed by the desire to extend human limits, to break barriers, to push the frontiers of physical and mental capability. They go out looking for a challenge and when they find it, they absorb it, they inhale it, it flows in their bloodstream. And then, when against all odds they manage to beat it out of their bloodstream, they go looking for more.

Jockeys are racing's elite athletes. On that one Saturday in November in the year 1993, Cheltenham's weighing room was exploding with the mighty force of the DNA of over thirty such elite athletes.

I was privileged to be one amongst them.

It seems only fitting at this point to explain the significance of Cheltenham to racing – the momentousness of the venue. This, in fact, sits at the heart of what is to follow in my story, it contextualizes why I did what I did, why I had to do it.

Cheltenham, very simply, is the Mecca of racing. I would ride there for nothing. It is what Wimbledon is to tennis or what the Tour de France is to cycling. It is the Olympics for horses and riders alike – where the best of the best ride to win. Nestled against the magnificent backdrop of Cleeve Hill, the 350-acre racecourse sits in a natural amphitheatre which over the course of the festival could attract as many as 200,000 spectators. It is unsurprising, therefore, the enormous sense of occasion one experiences at Cheltenham. Technically, it is a difficult course, highly demanding in the natural contours of the track, which makes it unlike any other racecourse in the country. For a jockey, it presents perhaps the greatest test of horse and rider – of the dynamic union of the two. This is why racing fans love Cheltenham; because, if performed well, what they witness is pure poetry in motion. So, for me, as for most jockeys, Cheltenham is the capstone venue for jump racing – a place where dreams are made, reputations are forged and history is written.

It is here, in Cheltenham, that this story is set.

It was the Saturday of The Open, the second Saturday of November. My first ride of the day was in the ironically named Murphy's Handicap Hurdle, a big prize-money race, on 2/1 favourite Arcot. It was the first time I was riding the horse, but I was there to win.

As the race progressed, Arcot and I travelled and jumped like a winning combination, my body following the rhythm of his stride pattern, the horse in tune with my cues. Then, at the second last hurdle, when the race was at its tightest, fastest and most competitive, Arcot made a costly error – he grabbed at the hurdle instead of jumping over it.

Balance was broken, momentum was not; our rhythm snapped short, and I was eating dirt.

The physical construction of hurdles – three and a half feet high, angled out to about seventy-five degrees – allows for horses to flick over them in the stride that they are galloping in. Fences are slightly different. About a foot and a half higher, horses have to get back in their hocks to jump them; they have to elevate themselves to clear a fence. Purely because of the physics of this – height, weight and momentum – statistically, the severity of a fall over a hurdle is much greater than one over a fence.

A mistake at a hurdle for good hurdlers usually involves a horse travelling at 35 mph, clipping the top of the hurdle, and falling over at 30 mph – there is nothing to break the momentum of that fall; you contend with 1,200 pounds of horse flesh at 30 mph on impact. Fences tend to be more forgiving, relatively speaking, of course. A mistake at a fence still involves a horse crashing into it at 35 mph. However, the height of the obstacle serves to break the momentum of the fall. When the horse falls, he falls over at a kinder speed – you only contend with 1,200 pounds of horse flesh at 10 or 15 mph, on impact. Piece of cake. In all seriousness, however, as a jockey, the fall that you fear most is at speed over hurdles.

This was exactly how I fell.

And if that wasn't enough, when I fell, I fell at the seminal second last.

In any National Hunt race, at any racecourse in the world, the second last hurdle takes on particular significance; riding over and past it involves arguably the most intense few seconds of the entire race. The second last serves as a crucial marker, where how you jump can determine whether you win or lose, where everyone is jockeying for position and no one is willing to give an inch. And there I was, in the middle of it all – in a winning position on the inside, twenty galloping horses around me vying for advantage – when my horse clipped the top of the hurdle. Such was the speed and intensity at which we were going that Arcot did a double somersault, fired me in the air and then crashed to a halt, throwing me to the ground like a spurned lover.

My prize from the first somersault was a kick in the head, so hard it split my helmet. My consolation prize from the second was a kick on my right wrist. Fully conscious on the ground, I curled up into a ball, as instinct dictates when we fall, but I was trapped in the concertina effect coming into the turn at the bend. Every horse that was behind me caught me in their stride, hooves striking the ground with as much as 3,000 pounds of force.

Prophetic? I didn't think so at the time.

Eventually, Arcot got to his feet. And so did I.

Was I in pain?

My God, I was in pain. I was white with pain.

But you don't give in to pain.

You get up and you get on with it.

I was lucid, I hadn't been knocked unconscious, the Mackeson Gold Cup was in thirty minutes and I was going to ride in it.

You do not give in to pain.

This I learnt in the most natural of ways when I was barely five or six years old. We were never mollycoddled as kids – you got on a pony, you fell off, you got back on again. Nobody came running to your aid, you just picked yourself up and got on with it. That was precisely what I was going to do on that day.

Arcot threw me off – it was bad luck, it was a bad fall, I was in bad shape, but it wasn't enough to deter me. I wanted to ride, I was going to ride. This wasn't any old racecourse. It was Cheltenham. This wasn't any old race. It was the Mackeson. And if I wanted to win it, I had to ride it. That's all there was to it. This was race-riding; there was no 'Play it again, Sam'. This was real time, the real deal. I had one shot and I wasn't going to give it up. The Mackeson Gold Cup was in thirty minutes and I was going to *win* it.

There was prestige in the Mackeson Gold Cup; it heralded the start of the season, the first big race. And if Cheltenham was the crown of jump racing, the Mackeson Gold Cup was one of the jewels in the crown. My teammate would be Bradbury Star, and with six wins from eight starts, his record at Cheltenham was unrivalled. More than that, this was personal; it was the confidence I had in my relationship with this horse. I had learnt right from my pony-racing days that no matter how skilled a jockey you may be, you cannot win if you don't have a horse that wants to race for you. Bradbury Star was a legend and I knew he wanted to race for me. Our dialogue was congruent; we were in harmony. I knew I could win.

I have always considered success in professional sports

– my sport or any sport – to be possible only if there is a contribution from the mind that is stronger than the contribution from the body. You have to be blessed with a very particular mindset where you 'feel' every situation, every curveball that gets thrown your way, and you challenge yourself to beat them all.

You have to see victory in the mind, you have to feel it, and you then have to believe it.

See it, feel it, believe it.

This was what fuelled my fire. This was what allowed me to ride as naturally and as fearlessly as I did. On the big day, I could perform. And this resolve only ever comes from the mind. If you haven't already executed it in the mind, the body will not follow through. I had already executed it in my mind. It was just my body that needed to follow through.

So I picked myself up, determined mind ignoring broken body. I was ready to ride.

After my fall, the correct protocol to follow would have been to get into one of the ambulances on site and be taken to the doctor in the medical room. I didn't follow protocol.

It's odd, you might think, but as a jockey riding a race, you don't see the ambulance following a hundred yards behind, of course you don't; you don't even imagine it. Yet every time you sit on a horse on a racecourse, it's there. (In fact, there are three of them: one in reserve, two following you around.) But you never even think about it, even though you come incredibly close to needing it. That day, I noticed it. And I needed it. And I deliberately avoided it.

I knew that if I had got into an ambulance full of paramedics, I would have had no chance of riding the next race;

they would have caught me out in a second. I was playing an elaborate game of wits; one wrong move and the game would be up in a heartbeat. I couldn't afford to get found out.

Instead, I got myself up off the ground and, as quickly as I could without being noticed, I jumped into a trailing maintenance jeep and asked to be taken to the medical room, allowing myself enough time to regain composure without being under medical observation. Once in the doctors' room, I pulled off the charade like an expert, managing to delude everyone – most of all myself. I mustered up enough wherewithal to answer all the questions they asked – what race was I riding, what had I ridden the day before, what day was it, what car was I driving – and somehow or other got myself passed fit to ride. I was not fit to ride. But I got away with it.

Before anyone had a chance to change their mind, I went quickly into the weighing room, changed into my colours and then I went to weigh out at the scales. By this point, if truth be told, I was in so much pain, I was losing focus. My brother Pat, standing by the weighing-room door, watched in total disbelief as I handed my saddle over to the trainer, Josh Gifford. My eyes were glazed over, I was staring into nothingness and frankly I had absolutely no idea who was taking the saddle from me – it could have been John Wayne for all I knew! Pat saw this clearly. Later he would tell me, 'I couldn't believe it. You just weren't with us. I promise you, you were not with us at all. And thinking, just for the minute, as a brother and not a jockey, as I watched you hand that saddle over, I said to myself – this isn't right, this just isn't right.'

This same voice of reason was echoed by my valet, Johnny Buckingham, when I went back into the weighing room to put my helmet on and have my cap tied. Halfway through tying my cap, he stopped, put his hands down by his side, looked at me and said, 'Ya, I'm not tying it, your eyes are not right.' The fact of the matter was that my eyes told the whole story, and in my eyes all these people could see the pain, they could see the distance; the eyes don't lie.

I tried to reason with him – I can be *very* articulate – but Johnny Buckingham would not budge one way and I would not budge the other. So, Johnny's assistant, Andy, who happened to be my friend, finally tied my cap because I had started tying it myself, so determined was I to ride. Johnny Buckingham would claim, years later, that he had never, ever seen anybody who was on a 'completely different planet' go out to ride in a race. And yet within me, I felt secure enough to pull off the task at hand.

Everybody told me not to do it – my brother, my valet, my girlfriend, my colleagues. In fact, everybody expected to hear an announcement of a change of jockey. But there I was. And as I sat on that horse in the parade ring, and cantered down to the start, I had already established my state of mind. Even though I was only semi-conscious, I was going to ride the most perfect race. So I went through the motions, almost on autopilot. I sat on my horse: heels down, toes up, knees bent, thighs tight, back straight, shoulders back, chin up, eyes forward, elbows bent, wrists in, fingers tight. And I was ready to start.

Every race can be won or lost at the start. Despite the fact that I knew I was not going to be contesting for the lead or even to be in the first quarter, I knew exactly how I was going to start. I jumped off as if I was going to make the running but all I was doing was ensuring that I was able to establish my horse's cruising speed quickly and not give away any ground in the process.

Inside my helmet, my head had started throbbing, the pain radiating from the base of my skull, with a grip so strong I had to take a sharp intake of breath. My out breath forced away the pain.

I was second in line to jump the first fence. This race had fourteen runners and by the time we had jumped one fence, the horses around me had dictated the speed they were going to run at. I got a feel for the pace of the race, and I knew what I needed to do. Approaching the second fence, the two front runners – Guiburn's Nephew and Egypt Mill Prince – had established the lead. I found myself shuffled back, in fourth position, riding at Bradbury Star's cruising speed, comfortable and confident. *Ride the race to suit the horse and not the horse to suit the race.*

From this point onwards, I had one goal and one goal only: to ensure that my horse got through this race as economically as possible without expending any more energy than he needed to. I knew that if I could sustain this, the horse would have the finish in him to win the race for me.

Tactically, it worked out beautifully as Jamie Osborne on Egypt Mill Prince in the lead had ensured a good gallop. *My head was thumping now, slowly and excruciatingly, a power tool grinding mechanically against my skull.* I noticed the horses around me jostling for position, afraid that the leader was going to get away from them. I was content where I was – in fact I was exactly where I wanted to be.

My position changed only as a result of the horses around me; I was very much still maintaining my pace, giving myself plenty of space at the fences.

The important thing riding over fences at Cheltenham is this: you need to get a horse to jump off its hocks and bend over the fences, you can't get in deep into those fences and get away with it too often, because they can be unforgiving. You need a horse to respond to your command, to be able to come up on the right stride, and to do that you have to be riding the right race. After the fourth fence, I knew I was riding the right race. I knew I was going to win.

As we approached the end of the first circuit, having ridden the first third of the race, I had dropped from fourth position to nearly seventh. *The grinder was now a hammer, sharp and heavy and unrelenting, banging away through the middle of my skull, forcing my eyes shut. I willed them open.* Halfway down the back straight, I could see Jamie Osborne up front, tactically trying to fill his horse's lungs with air. He slowed down to do this, and being the leader, in doing so, he eased the pace of the race. Because I had maintained my horse's cruising speed throughout, this now worked to my advantage – the others had slowed while I had stayed steady. Exactly as I had predicted, this propelled me forward.

Going into the last open ditch, four from home, Jamie Osborne had picked up speed again. *The pain flashed hot and hard and out of control – almost, out of control.* The horses behind Egypt Mill Prince were desperately trying to close in, and again I found myself pushed back to fifth position, convinced completely by the decision I had made to hold my pace right from flag-fall.

It was a chainsaw now, slicing into my temples. The blood roared behind my ears; my vision blurred.

The contours of the track at Cheltenham can affect jockeys as much as they can affect horses – you get to ride past the grandstand, not once but twice in every race. And even when you're as far as halfway down the far side, you can hear the cheers of the crowd. Very often this can get jockeys to commit sooner than they should; it's easy to get caught up in the wave of exuberance. That's another reason why Cheltenham is different to any other racecourse, that's why it's so demanding. It was a classic mistake and I was not going to make it.

As we started to approach the brow of the hill, the race was being run at a furious pace and I could see considerable jockeying for position among the first four or five runners. Then, just before the third last fence, I deliberately manoeuvred Bradbury Star to the centre of the course coming down the hill. I knew the ground fell away at the back of the fences on the inside and the outside, and I knew how dangerous this could be for me. Second Schedual followed exactly the path I had tried to avoid, and when he fell, I knew it was my chance to pounce.

Inside, my head was pounding with an intensity so great that it was almost pushing my helmet off. As if my brain was trying to escape but couldn't find its way out. I willed myself, willed every last atom in my being, to concentrate. I landed running at the back of the second last fence. Egypt Mill Prince had quickened again up front. I slotted in behind him with one last fence to jump.

By now, I was riding on adrenalin and adrenalin alone. Turning into the straight, seeing Jamie Osborne on Egypt

Mill Prince directly in front of me, I focused my gaze between my horse's ears. *I could feel the rhythm of blood pulsating inside my head. I winced, my brow furrowed, tight, not with pain, but with resolve. I was on the brink of collapse, but I couldn't let that happen. I wouldn't let that happen.* My mind converged on the eye of the storm. I was a horseman riding to win.

It was at this point that my eyes began to give way. I started to see two Jamie Osbornes. But, as things stood, I was going to beat them both.

Bradbury Star kept in a straight line, and with remarkable ease, cruised into the lead halfway up the run-in. He pricked his ears forward, loving every moment of those last victorious strides. *There was an explosion in my head as if my skull had burst; as if my tortured brain had finally found a way out. With whatever little I had left in me, I continued to focus my eyes between Bradbury Star's ears, drowning out the waves of crippling pain that threatened to swallow me up and eat me alive.* My horse ran all the way to the line. We won by seven lengths. He had done it. I had done it. We had done it.

No safe. No pause. No predictable.

Crushing pain. Blinding headache. Double vision.

No retake. No rain break. No sitting out. No backing down.

See it, feel it, believe it. Faster. Higher. Stronger.

Looking back . . .

There was no good reason for me to have done what I did. In fact, it was nothing short of sheer insanity. Because the moment I passed the winning post, the adrenalin stopped. Just like that. I had been on the brink of consciousness, riding on the edge, and now that edge gave way. For the past seven minutes, my mind had been way ahead of my body; now the two equalized, giving way to crushing, indescribable agony. My legs buckled as I dismounted, and after that I could do no more. Even the walk to the weighing room seemed an insurmountable task and I had to be supported to the scales by my girlfriend, Joanna, and Althea, Josh Gifford's wife, while I clutched my saddle and lead cloth, my teeth clenched in pain. If you had asked me, at that point, what race I'd run in, which horse I'd ridden, I wouldn't have had the faintest idea. In the weighing room, I collapsed.

So why did I do it? Because I was desperate to win the Mackeson at all costs. Athletes are like that when they have an ambition – they are consumed by an unquenchable desire to fulfil that ambition, no matter how unachievable it may seem. *I* was like that. For me, the power of my mind, the power to put mind over matter, the power to position myself where I needed to be in my own head – that is what kept me going. But I wasn't some adrenalin junkie on a suffer-fest. This was no swashbuckling display of valour. It was an intelligent, calculated decision on my part. The second I got myself on my feet after my fall, I had reasoned it through. I had weighed the risks and considered the odds, and I knew the rewards would be well worth it. And so there was nothing that was going to keep me from missing my chance. Especially not pain.

I believe that all of us, all jockeys, have an understanding with pain. So in control do we feel in the environment that we are in, we simply don't see pain as it is perceived by others. This is the secret language between us; so we laugh off the broken collarbones and the concussed heads, the fractured wrists and the bruised ribs. We get up and we get on.

It takes patience, strength, guts, determination to ride a racehorse. Courage. Tears, tantrums and everything else. So we deny physical pain by not recognizing it as such. Instead, we view it as an encumbrance. And when our adrenalin is pumping, when we are in this state of mind, we create an aura around ourselves that rejects encumbrance. We *have* to believe we are unbreakable. This is the state of mind one is in, as a competitive sportsperson. This is the state of mind one *needs* to be in, as a competitive sportsperson.

So, getting up to ride in the next race was one thing. It was getting up after that fall that was something else. Falls are falls and, in general, with a fall, in nine out of ten cases, you're going to walk away absolutely fine, and never get touched by your own horse or any other horse. But two-thirds of the field in that race at Cheltenham had kicked me on the ground like I was a football. My condition was diagnosed as 'general exhaustion' but it was delayed concussion that I had really suffered.

Whether or not I was right in riding the race in the condition that I was in, I don't know. But I had started something and I had to finish it.

How did I do it? I used my head.

The art of riding at any racetrack is understanding that there is only one winning post. A good jockey understands the subtle nuances of the tracks at different racecourses. At Cheltenham, races can develop earlier as a consequence of the contours, as a consequence of running off the top of the hill, which is still a long way from home. Because of this, there is a tendency for the jockey to get caught up in the momentum too soon and his horse ends up running the race as if the last hurdle is the winning post. This means, by default, that the horse is not doing its best work at the finish. In my opinion, to ride Cheltenham well you have to have a horse running all the way to the line. To have a horse running all the way to the line, you have to have ridden a perfect race.

Horses are naturally competitive animals, but they can become disillusioned if you burn them out too quickly. The trick is to fuel them, to give them a ride so they have enough left in them to channel that competitiveness where it counts. This was my overriding objective with Bradbury Star, and I knew that he would respond. I was betting on the fact that he would have enough left to give when I most needed it, because to win in the condition I was in, I depended heavily on the energy I had reserved. Had he needed any extra assistance from me, I would not have been in a position to give it. It was a leap of faith; I took it.

A lot of what you do in life is what you're born with.

To everybody else around, I might have been looking into space. And whether or not I saw myself as a jockey or wanted to make a life from it, there was obviously someone inside me that rides races, rides racehorses, that was working away the whole time.

My ride made great copy. The next day there were news articles in the sports sections of every newspaper in the country extolling my audacity, my courage and my resolve. Colleagues, trainers and journalists alike expressed unreserved admiration for this feat I had pulled off against the odds. I was called 'The Ironman', 'The Hero of Prestbury Park', 'Lazarus of Bethany' and other such hyperbolic names. I won the Mackeson Meeting 'Man of the Match' award, judged by former champion Terry Biddlecombe, for the best riding performance over the two days at Cheltenham.

My star was burning bright.

I felt invincible.

But pride, as they say, comes before a fall.

Little did I know the fate that awaited me.

Little did I know that everything that I believed in about pain and control and courage would soon come crumbling down.

Little did I know that Arcot wasn't quite done with me.

Horse and Pony

Kilteely, Thurles, 1981.

No matter how many horses I've ridden in my life, and how many races I've won, my first time riding point-to-point will be indelibly imprinted in my mind. It was a life lesson – but not for the reasons you would expect.

My Uncle Mikey lived on our grandfather's farm in Kilfrush – this is where my father was brought up – about 3 miles from Bank Place. We kept our cows at Uncle Mikey's and we would often go and milk them by hand before and after school. Mikey also had horses on his farm and, being a formidable horseman himself, he would always encourage us kids to ride. And so the avid riders among us – Laurence, Pat, Eamon, Kathleen and I – would often go and ride out at Kilfrush.

It is easy to remember exceptional horses, just as it is to remember exceptional people. When I was about fifteen years old, Uncle Mikey had two exceptionally good horses – a mare called Luck Daughter and a gelding called Kilteely.

It so happened that he also had an exceptionally strong addiction to the bottle.

Uncle Mikey would alternate between periods of total inebriation and complete sobriety, and in my eyes, I had two uncles.

When he was drunk – for two or three weeks at a time – he was the most irresponsible man you could find. He would neglect everything: his house, his horses, his farm; nothing mattered. Uncle Mikey Hyde.

When he was sober – for a month or two months, maybe three months – he was the most conscientious man you could find; a brilliant horseman, an impeccable farmer. Uncle Mikey Jekyll.

Just before my fifteenth birthday, Uncle Mikey was going to let me have a ride on Kilteely in my first-ever point-to-point race. I was terribly excited to try something

I hadn't done before, and I waited in eager anticipation for the day to arrive. On the morning of the race, however, when my father drove me to Kilfrush, I discovered to my dismay that my uncle had found his bottle the day before. He was in bed, dead to the world. The horses hadn't been fed, the stables hadn't been cleaned; the house, the farm, the animals lay in a state of neglect. Everything was in complete disarray.

The wave of disappointment that washed over me was powerful. And then pragmatism took over. Both the problem and the solution lay before me. If I wanted to race, I had to take control of a situation that had spiralled out of control. So I fed the horses, I cleaned the stables, I prepared Kilteely for the race, and put him in the horsebox myself.

Richard Shanahan, who worked for a local trainer, Joe Crow, and rode some of Mikey's horses, helped me to drag my uncle out of bed and into the car. Richard then drove the car, with the horse in the trailer. I sat in the back and next to Richard, in the passenger seat, was my uncle, asleep, drunk.

On the way to the races at Thurles, Uncle Mikey woke up and decided that he had to stop at the pub in the village of Lattin. We tried to dissuade him, pleading punctuality, but what chance does logic stand against the maniacal urge of an addiction? Our arguments fell on deaf ears. Mikey promised not to stay long. We relented.

An hour later, we still couldn't get Mikey out of the pub.

Once again, I felt that familiar wave of disappointment. But once again, a cool head prevailed. I was laid-back by nature, but I was also determined. I had a definite stubborn streak to me and when I wanted to do something, I usually found a way to make it happen. If we couldn't get Mikey out of the pub, we couldn't get Mikey out of the pub. But if I wanted to race at Thurles, I needed to be at Thurles. I wasn't going to give it up so easily. So we decided to go without him. Richard Shanahan drove Kilteely and me to the races. I saddled the horse myself, weighed out on my own and I rode my first point-to-point on Kilteely while Uncle Mikey – of his own making – missed the opportunity of seeing both horse and nephew run the kind of race that he would have been eminently proud of.

This incident stayed with me for a very long time. But it wasn't the ride or the win or the thrill that I remember most – it was the fact that my uncle had been too busy drinking himself into a mad stupor to care. I understood then how easy it is for a person to lose control, to let their weaknesses destroy them. I vowed never to let this happen to me.

In a lighter vein, I'd like to narrate a very different, but equally significant, story from those early years. Rewind to 1979, during which time I could still do the weight required for pony racing. The protagonist of this tale is a little grey pony called Bula Lady. I held the champion pony rider's title at the time, and shortly after our experience below, my sister Kathleen, a supremely gifted rider, would win the title from me. Indeed, Bula Lady was Kathleen's pony and my story is a tribute to her.

For me, one of the greatest and most unexpected surprises to come out of the pony-racing days was the time I got to spend with my father – just him and me – as he drove me back and forth from the various racecourses.

One always underestimates the time one has left with one's parents, and these were the moments to be cherished – the good times, the heart to hearts, the laughs.

'The mystery of the missing pony', for example, is one such childhood story that Kathleen and I still reminisce fondly over. I was thirteen years old at the time, Kathleen was twelve. We were driving back home, having spent a long day pony racing in Galway. Dad was driving, I was asleep in the front of the car, Kathleen and our friend Richie Farrell were asleep in the back of the car, and in the horsebox behind the car was Bula Lady. The journey from Galway to our home was a distance of approximately 80 miles, and dusk had fallen as we turned into 5 Bank Place. Just as Dad was parking the car in the drive, Kathleen woke up and looked behind her. Then she did a classic double-take. *Then* she patted our father on the shoulder and said, 'Where's the pony?'

I will never forget the look on my father's face when he turned around that day. Engine still running, he jumped out of the car in a blind panic, but of course Kathleen was right – there wasn't a horsebox in sight!

Oh, the state of fluster everyone was in!

I have always had a love of fast cars and fast driving, but even after all these years, I don't think anyone drove as fast as my father did that evening. Within a split second, he had reversed feverishly out of the driveway and sped out on to the road. The quest for the missing pony had begun. Four heads turned this way and that as we retraced our steps, expecting that the trailer would have come off not far from home. Instead, we found ourselves some 10 miles back, all the way in Caherconlish. There, in a small park off the main road, was the horsebox, lying on its side, while Bula Lady, delighted with her lot, was being walked around by a group of young boys.

We learnt from them that the horsebox had come off the car as we drove up through the high street, then slid out between the two cars behind us and gone right through the entrance of the park, all the way in, without so much as a scratch. Then the towbar of the horsebox had hit a tree and turned it over. The guys had opened it up and led the pony out. Bula Lady was unscathed, none the wiser, and two carrots richer!

1980. The torch passes. Kathleen winning the champion pony rider title from me.

The four of us drove home – with the pony safely in the horsebox and the horsebox safely behind the car – laughing about 'the mystery of the missing pony' the entire way back. An insignificant blip in the story of life, but one that we would remember decades later. These were the great times, the stuff that sticks, the memories, the sound of our laughter . . .

I Soar

'When I bestride him, I soar, I am a hawk; he trots the air. The earth sings when he touches it.'

William Shakespeare, *Henry V*

That Place in Time

There is an attitude to winning.

I think after all the falls and scrapes, defeats and near wins, blood and bruises – to my body and my ego both – I had finally learnt this. There is a definite attitude to winning. And this comes only from experience, from confidence, from the taste of having won before. This is what propels you to win again. It inspires you, drives you, pushes you – almost to madness. And you don't rest until you quench those fires of desire. Until you reach perfection. Until you ride the best race of your life. Then, you breathe. There is a stillness, then. A state of nirvana.

This is where I was, in this state of absolute bliss, on 27 April 1994.

I was at the last fence at Cheltenham, riding Gale Again for Tommy Stack in the Silver Trophy Chase, fighting to win the race of my life.

Five days later, I would find myself in a different place, fighting to win a different race. This time I would be at the last hurdle at Haydock Park, riding Arcot for Jeremy Glover in the Crowther Homes Swinton Hurdle, fighting to win the race *for* my life.

And so Gale Again would be my last horse, in my last race before my life changed for ever. Before the end.

But I take heart. Because every story has a beginning and an end, and if mine had to end, this is how I would have liked it to end – to be riding a horse who knew how to race, for a man who truly understood racing.

Before the end, however, comes the beginning.

The first time I ever saw – just saw, not met – Tommy Stack was at the Thomastown Castle Stud in 1979. I was thirteen years old at the time, working weekends at the quarantine with Francis O'Callaghan. I had accompanied Francis to help organize the shipment of horses from Tommy Stack's stud that were due to fly abroad. Francis was responsible for their transportation from the stud to his own yard, where they would join the quarantine. Then, at the end of the quarantine, the horses would be transported from Francis's yard to the airport. When Francis had told me who we were going to see, I had been unable to contain my excitement. And then, when I finally saw him, I stopped and stared, and stared and stopped because I couldn't believe I was looking at him in the flesh: Tommy Stack, *the* Tommy Stack of Red Rum fame, was standing some ten metres away from me.

Some encounters only last seconds but stay with you for ever; this was one of them.

A decade later, in January 1989, at the time I was riding for Barney Curley, our paths would unexpectedly cross again. I knew that Barney Curley and Tommy Stack were good friends from Barney's Ireland days, and, to my delight, Tommy Stack asked Barney Curley if I would ride his horse in the big race in Ireland in a few days' time. It was a huge honour for any jockey in my position – Tommy Stack was one of Ireland's most widely respected trainers; he had been Champion Jockey, he'd won the Grand National, and Red Rum, who he had won the National on, was perhaps the best-known racehorse of all time in UK racing history. In my eyes, the man was one of the greats. I was elated.

I was to ride at Leopardstown in the Ladbroke Handicap Hurdle, one of the most competitive hurdle races of the year in Ireland. The horse I was to ride was Tommy Stack's only jumper at the time, a big, tall, rangy, unfurnished gelding called Kingsmill.

The night before the race, the ex-champion-turned-trainer and I met up in my hotel and sat down to discuss tactics on how best to ride Kingsmill. It was not the most straightforward of situations – there was a huge punt on the horse – but having only run a couple of times before, he was still very inexperienced; essentially, a novice in a big handicap. And so we discussed strategies on the optimal way to mask his inexperience and get him to deliver.

Come race day, however, despite my best efforts, he finished third to Redundant Pal and Atteses. I realized after the fact what I could have done better. Every horse is different and it is very important in racing to factor this in, and then to understand, as partners, how much the two of you can give of yourselves, for as long as you can give and to where it will take you. With Kingsmill, it was all about rhythm and I had ridden him too forward in the race, a style which, in hindsight, didn't do him justice. When I came back in, I remember telling Tommy Stack, 'If I rode that horse again, I'd ride him with different tactics.'

It so happened that I did get the opportunity to ride Kingsmill again, at Leopardstown in the Irish Champion Hurdle, one month later on 4 February 1989. As in the Ladbroke Hurdle, he was once again the gamble of the race, being backed from 14/1 to 8/1. Both Kingsmill and I had learnt from riding Leopardstown a month previously, and I knew what to do differently this time.

As soon as the race started, I put it to the test. I held up Kingsmill as Master Swordsman set a fast pace, establishing a five-length advantage from an early stage, followed by Paddy Mullins's Cloughtaney. For horses of that class, I felt that our rivals seemed to be going too fast early on and that it strung them out. In my opinion, there was only one way to ride Leopardstown and that was around the inner. So I decided to take a chance and drop Kingsmill out and let him jump and creep his way into contention when the leaders began to tie up. I was deliberately riding a waiting race and I was able to establish Kingsmill's cruising speed, which I didn't feel I had done effectively in the Ladbroke Handicap. The rain had loosened the ground and that also helped him.

After the third last, Cloughtaney took over from Master Swordsman, chased by Jim Bolger's Elementary. I went third into the straight with Kingsmill jumping beautifully, travelling beautifully. It may have looked as though Kingsmill was the only one to quicken in the straight but it was more a case of keeping up the gallop as the rest faltered. This was a calculated piece of riding on my part and when I dropped Kingsmill in, making gradual progress from the fourth last, he was travelling easily.

We cruised up to take the lead approaching the last, and

then it was over: in a dramatic improvement to his performance from just a month ago, he won the Irish Champion Hurdle by ten lengths. I delivered the 8/1 chance without resorting to the stick.

It was my first big winner for Tommy Stack. In a post-race interview, he would say, 'I'm very impressed by Declan – not only by his riding but by his attitude before and after a race. He puts a lot of thought into what he's going to do and his post-race explanations are first class. He should go a long way.'

Our trust and our friendship both began to blossom from that moment on.

He started bringing me over to Ireland regularly to ride his horses, and sometimes he would send his horses over to England when the ground in Ireland was too soft. One of the most outstanding horses he had at the time, that ran in England, was Gale Again.

I first had the pleasure of 'meeting' Gale Again at Sandown Park in the last week of March 1994. He was a black gelding, not more than 16 hands tall, with a short back, and a beautiful stride low to the ground, showing a real preference for fast ground. I had never ridden the horse, and although Tommy Stack had briefed me on him prior to the race, I was in for a wonderful treat. Because when I actually rode him for the first time, he stayed true to his name – he quite forcefully blew me away. I remember being so surprised at how electric he was over an obstacle; so quick, so efficient, so economical. He excited me no end and I remember thinking to myself, If I could ever create that perfect horse to be partnered by me, it would be in the image of Gale Again.

A victory so easily achieved at Sandown led to high aspirations for Gale Again. Tommy Stack decided to run him in the Grand National the following month. I was cautiously optimistic in the run-up to the day. I felt that with good ground, there was no reason why this horse wouldn't stand a great chance of winning the event – he was such a class act, among the best of the best.

Regrettably, the weather gods conspired against us. Come National day, the heavens opened in Liverpool, and Gale Again was taken out, quite wisely, by Tommy Stack, for soft ground. He went instead to the intense racing theatre that is Cheltenham, where I rode him in the Silver Trophy Chase a few weeks later.

The thrill of riding a horse with such tremendous pace and the ability to jump as fast and as elegantly as Gale Again could is second to none, and I consider myself fortunate to have ridden him over the course of my career as a professional jockey. That it would turn out to be both the best, and the last, race of my life – as I knew it – makes it all the more poignant.

The Silver Trophy Chase is run over 2 miles and 5 furlongs and is usually the highlight of the first day of The April Meeting at Cheltenham. That year, the race was being run with six runners, all good horses, making it a very competitive field. It's always great to line up in a race where you know there will be a contested pace, and Wind Force ridden by Neale Doughty for Gordon Richards was a guaranteed pacesetter, as was Garrison Savannah, ridden by Richard Dunwoody for Jenny Pitman. For me, this meant I could slot my horse into his cruising speed quite quickly and ride a patient race. We had seventeen fences to jump in all and this was exactly the kind of race where I could ride the race to suit my horse, very easily, very effectively.

Right from the start and through the top half of the race, Wind Force and Garrison Savannah were sustaining a good gallop. They were followed by Far Senior and Southern Minstrel. Graham McCourt on Elfast was riding, like me, a waiting race and I knew he would prove to be my biggest danger. To my advantage, he was tracking the field, riding tactically, and I slotted in behind him. My double advantage was that for a lot of the race he seemed to forget that I was behind him – I capitalized on this small lapse in his judgement.

As we set out on the final circuit, the next was a downhill fence, the eighth. It was still Wind Force that had made every yard so far, leading Garrison Savannah by a couple of lengths. Keeping tabs on Garrison Savannah was Southern Minstrel towards the outside, then Far Senior a couple of lengths down in fourth place. Then Elfast with Graham McCourt in fifth, and yet to make any sort of move, to the back of the field, I was patiently riding Gale Again.

Below me, Gale Again was at his glorious best – the dream horse that every race-rider wishes the genie would grant him. So perfect was his stride, when in rhythm with it, you felt you couldn't meet a fence wrong. *Hind leg, hind leg, fore leg, fore leg* – the dance of his gallop, the beat of our bodies, as we raced together, man and beast, as one. To have that level of trust in a horse and to have that horse reciprocate with trust in you is exhilarating, a feeling second to none. You ride to win, of course, you always ride to win. But in cases like this, you also ride for the joy of riding. You feel an intoxicating sense of freedom. You don't want to be anywhere else, doing anything else. You feel – alive.

Six from home, I gradually started moving closer. Four from home, I had crept up to third place. As we raced down towards the third last, Wind Force was still in front, Elfast was stalking Wind Force and I was biding my time, waiting for them to play their hand.

When we rounded the final bend with two fences to jump, Wind Force seemed to be leading only on sufferance. Behind him, Graham McCourt was brilliant on Elfast, as savvy as a rider could get. At this point, for all intents and purposes, it was only him and me in the race, playing a game of poker on horseback. It was just a matter of who would outsmart whom, and when.

Then, just as Graham McCourt kicked on smoothly upsides Wind Force, thinking he would slip away from me, I stayed my pace, still directly in his shadow. He looked over his shoulder at this point to see the danger behind him, but he couldn't see me, such was my position – my position to pounce.

But I still wasn't ready to play my trump card.

In any race, you have to know the horse you are riding, you have to have established his cruising speed, you have to know exactly what that horse will find from the gear you are travelling in. And you have to know when to release that.

I released Gale Again getting away from the back of the second last.

And then it was Elfast and Gale Again stride for stride as we raced towards the seventeenth and final fence, and

I produced Gale Again to challenge at the last. I had total belief in him when I fired him at this last fence. He sprang off the ground from outside the wing as if he was a bird in flight. Such was the intensity of that final mighty fight for the finish that, had either one of us fallen at the fence, we would still be falling . . .

We ended twenty-five lengths clear of Far Senior in third place.

I'm not sure I've ever seen a race where two horses have gone to that last fence at Cheltenham at the speed that those two horses went. Any spectator watching this race wouldn't have had the heart to take sides. On display were two jockeys riding as well as they could ride.

A race of winners.

A true performance.

Many years later, Tommy Stack would recount the details of this particular race with both great nostalgia and great pride. 'Declan is the one person you want on your horse in any race,' he would say, his voice brimming over with genuine emotion.

And so, when I look back on this race, I feel a deep sense of fulfilment. And achievement. And joy. But also, sadness. Undoubtedly there is a tinge of sadness. Because if the person riding that horse was me, then I think I had reached a place in time on horseback that very few people get to. To be so at one with a horse, to be able to sit on a horse and travel its speed and jump like that and always be completely in rhythm with that horse, that is where I was on that ride.

And *maybe*, just maybe, this was the moment – that

moment of imperturbable calm when the fires of desire have finally been quenched – when I was just beginning to *want* to be a jockey. Maybe I had reached a point, maybe I had taken it to a place that I didn't expect to take it to . . .

People say it is easy to ride good horses, but to ride a horse in a way in which you are virtually controlling the entire event is a different matter altogether. I had achieved the ultimate goal. I had ridden the best race of my life.

Would there be better?

I would never know.

Five days later, on Bank Holiday Monday, in the direct line of sight of a grandstand full of holidaymakers, I would fall off a horse at the last hurdle.

A typical fall off a horse is no more than eight feet to the ground.

That's how far my body fell.

What they don't measure is how far the mind falls.

My mind fell down a black hole.

Deep and dark.

The wreckage could never be found.

But on this, the twenty-seventh day of April 1994, in the Cheltenham Silver Trophy Chase on Gale Again, I couldn't be further from the bottom. This particular ride on this particular day on this particular horse, I was in a place in time I didn't expect that I would ever be. The view from the peak was magnificent.

Christopher Goulding had once said of me that I have the appearance of being born astride a horse. If there was ever a time to live out his words, it was this.

This place in time.

Hello

Irish jump jockey Declan Murphy, 27, is off ▶ the life-support machine. The jockey was rushed unconscious to hospital in Liverpool, after he was kicked in the head during a horrific fall at Haydock Park. Last week his condition was described as "still seriously ill but improving". The County Limerick-born jockey underwent surgery to remove a blood clot on the brain. Declan had been riding the race's favourite, Arcot, when the pair fell at the final flight.

Brigitte Bardot spent May Day sitting on a sand dune in the Medoc region near Bordeaux, but she wasn't sunbathing. She was making her traditional protest against the slaughter of turtledoves. Hunting the bird was outlawed in France in 1969, and banned by a 1979 European Community ruling. With a battalion of ecologists, she waited for the annual shoot to begin. Shots ▼ could be heard in the distance, but the hunters stayed away from Brigitte.

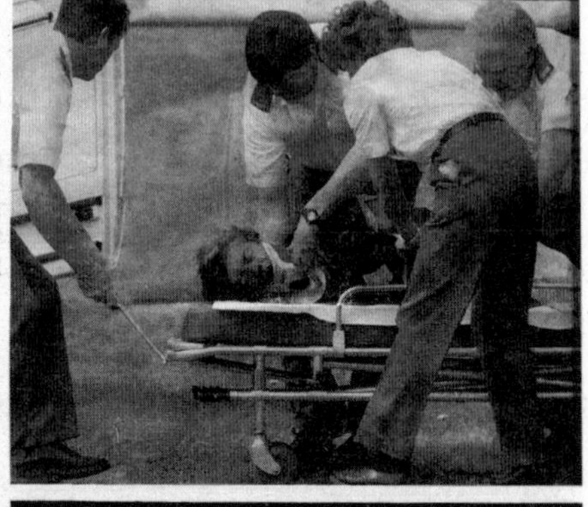

PANORAMA

◀ President Bill Clinton has a pow-wow with an unidentified native American tribal leader at the White House. The President shook hands with the tribesman, who was dressed in traditional costume. It was a meeting with leaders of many of the US's native American tribes and, according to White House officials, the gathering was the first time all of the nation's tribes were invited to meet the President.

▲ With the death toll already standing at nearly a quarter of a million people in Rwanda, desperate crowds of refugees have been waiting for rations of beans and grain. Chaos and violence have ruled since the president was killed four weeks ago and the latest horror was the brutal killing of 21 orphans and 13 Red Cross volunteers. The UN is trying to bring the rival factions together for talks.

'The reports of my death have been greatly exaggerated.'

Mark Twain

Penumbra

Monday, 2 May 1994

It was all over before it even began.

The race had ended.

The dust had settled.

The air was still.

The racetrack stood eerily deserted.

Arcot lay on the ground, destroyed.

Next to him, I lay motionless.

Then, as if possessed, the left side of my body started convulsing uncontrollably. One arm, one leg, one side of my torso lifted themselves up off the ground and fell back again in rapid, violent jerks. The other arm, the other leg, the other side of my torso rested in blissful ignorance.

The signals generated by my damaged brain darted around dangerously as the two sides of my body lost equilibrium.

But even as the blood spurted from my brain and out of every orifice in my head, my eyes remained closed, peacefully, like those of a sleeping child.

Then, the paramedics arrived. To exorcize me.

It takes five minutes to boil an egg. It took four minutes to decide if I lived or died.

With the speed and efficiency of experts, the paramedics on site identified that I was having a seizure caused by bleeding from the brain. Look at that photograph of me in the global 'Panorama' section of *Hello!* magazine – sharing page space with Brigitte Bardot and Bill Clinton – lying on the stretcher in a neck brace, eyes closed, nose and mouth gurgling with blood. One of the paramedics has his hand under the right side of my head, along the length of my face, trying to stem the flow of blood from my brain. I never knew this man, nor would I ever meet him again, but these are the unsung heroes of one's life. Whoever he is and wherever he is, I know now that his expertise in knowing exactly what to do at that moment in time was what kept me alive. Even if just for that moment. In the next moment, everything could have changed, such was my state of flux. But it didn't matter. Every moment that I wasn't dead, I was still alive.

From Haydock Park, I was transferred swiftly into the waiting ambulance and rushed to Warrington General Hospital. At Warrington, I had the second of my two seizures. A team of experts put me on a ventilator and stabilized my vitals so that I could live long enough to be moved to The Walton Centre for Neurology. They had one goal and one goal only – to keep me alive through the journey.

Later, I was told how fortunate I had been: had I arrived at Warrington four minutes later, I would have died. Four minutes; not three, not five. How it is possible to estimate someone's chances like that, to narrow down their window of life or death with the precision of a clock, boggles the mind. And yet, they could; it was science.

When I met Charlie Whittingham for the first time, in the summer of 1985, he paid me one of the biggest compliments of my career. He said I could ride 4 furlongs on a horse in 11 seconds, 11 seconds, 11 seconds, 10 seconds – that I had a clock in my head. How I was able to do this, to break down the pace so precisely, I cannot explain, but I knew the fractions completely by instinct. This was not something that you could teach somebody, it came from within. You either had it or you didn't; it was art.

My art, their science.

Which one was greater?

I question it not.

The answer is too humbling.

The Walton Centre for Neurology and Neurosurgery, part of the NHS Foundation Trust, is an institutional-looking building, every bit as sombre in appearance as the business it conducts inside its red-brick and algae-green walls. At the time of my accident, it was one of only two hospitals in the country equipped to deal with brain trauma to the extent I had suffered; the other being Southampton, 240 miles away – as far south of England as Liverpool was north. To my great luck, The Walton Centre happened to be located a mere 26 miles from Haydock Park. It was there that I arrived, on a ventilator, in an ambulance, under police escort – such was the urgency of 'now'. It was there that an innocuous-looking Welshman would be summoned from the golf course to save my life.

I wouldn't meet Professor John Miles until much later,

until after he had operated on me, until after I came out of my coma, but if you asked me to name the people without whom I couldn't have survived, he'd rank top of the list. I cannot think of him or say his name without feeling immense gratitude towards him, for his skill, for his science. Perhaps he was just doing his job, living by the Hippocratic Oath, fulfilling a call of duty, but for me, it was enough to build faith. I put my life in his hands. No, actually *my life was put* in his hands. And he saved it. That's faith.

When Professor Miles got the call, he was in the midst of a far more hazardous undertaking than trying to save lives. He was on the 5th hole at Heswall Golf Club, Wirral, trying to save par. Over the phone he was apprised of my condition from the CT scan that accompanied me from Warrington – it showed multiple fractures to my base-of-skull bone, and two blood clots on the inside of my skull, close to my brain. These clots were taking up space in my skull and, in so doing, were squeezing down on my brain. Because the brain – a soft cheese-like organ – is enclosed within the rigid skull, there is danger of compression caused by leaking blood. With enough bleeding, there can be so much build-up of pressure that oxygen-rich blood is prevented from flowing into the brain tissue, causing the brain to swell. Knowing from his experience that it would take at least twenty minutes to raise the bone flap sufficiently to give access to the blood clots, it was his decision to get the operation started straight away, before he even arrived on the scene. You cannot reverse damage to the brain – the clots had to be removed quickly.

The skill of a surgeon is determined not only by their dexterity with the knife, but also by their ability to recognize when there is a sense of urgency, when something needs doing straight away. Years later, Professor Miles would tell me himself that even highly qualified neurosurgeons sometimes lack the constitution to do this, for one simple reason: it requires them to take chances. But it is this ability – this courage – to make these decisions that saves lives. And when I look back, there is no doubt about it – Professor Miles not only had the brains to save my life, he had the stomach.

And so another phone call was made, this time from Professor Miles to Professor Rosalind Mitchell, who was on site at The Walton Centre. It was under Professor Mitchell's supervision that I was prepped for emergency brain surgery. My hair was cut, the right side of my head was shaved, and the points where the incisions would be made to cut open my head were marked out by tiny coloured dots.

I was in the operating theatre throughout the night of 2 May. During this time, the surgeons separated my scalp, made little holes in my skull with a sawing wire, and cut between the holes. Then, they lifted the bone flap. Much like one opens a hinged door, one side was lifted while the other side remained attached to the temporal muscles. Once the doorway into my head was created, they had the access they needed to go in and suck away my clots.

The scans showed the position of the clots clearly. The first one lay between my skull and the dura – the outermost, toughest and most fibrous of the three membranes covering the brain and spinal cord.

The second, more dangerous, clot was inside the dura, against my brain.

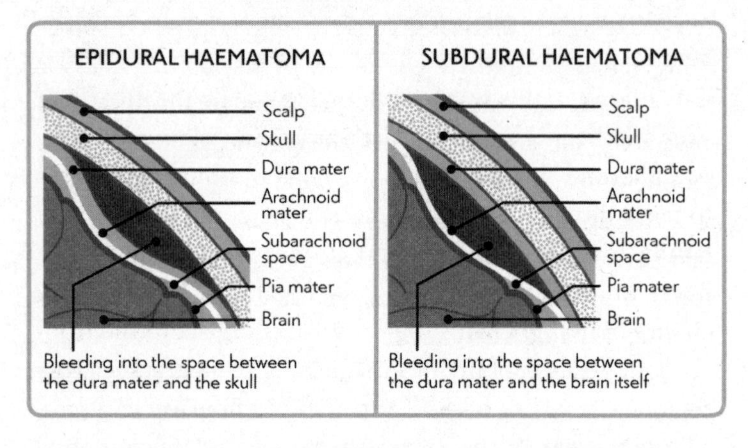

Source: BrainLine.org

Because of the position of the clots, it was inevitable that I would suffer significant nerve damage during the course of the operation. But it was the clots that Professor Miles and his team chose to focus their attention on – everything else became collateral damage, the unavoidable casualties of war; the clots *were* the enemy. So, a split-second decision was made to keep me alive at the risk of damaging the nerves, knowing that it was a near certainty that this would leave me paralysed and partially blind.

The dura duly opened up, the two incriminating blood clots inside my head were suctioned out; first the outer, then the inner. Before signing off, several tubes were placed inside my head to allow for drainage in case any new blood vessels had ruptured as a consequence of the operation – if left unattended, these could take me back to square one, creating new clots, which in turn could cause my brain to swell. There wasn't a single wasted moment nor a single wasted movement; once the job was done, my head was sewn back up. The operation was over. Now the waiting game began.

While all of this was happening inside the theatre, there was another, equally dramatic scene unfolding outside. After several hours of receiving non-committal responses to fairly direct questions, such as 'Is he going to live?', my family was given the prognosis. If I survived the three hours immediately following my operation, I had a 50:50 chance of living. I had suffered a direct impact to the frontal right position of my head, which had sent cracks across the bone; my base-of-skull was fractured in twelve places. The proximity of the cranial nerves, especially my optic nerve, posed a grave danger to me – the doctors thought I might lose my hearing, have impaired speech and were almost certain that I would lose vision completely in my right eye.

All of this, of course, was only pertinent if my brain would react. If it didn't, we'd have to, in medical slang, *C/C* – *'Cancel Christmas'* (dead). In other words:

a) My chances of surviving even three hours were speculative, at best.
b) If I was still alive in 181 minutes, whether I'd be able to sustain that for much longer was a coin flip.
c) Heads or tails? If the coin landed the right way up, there were no bets on the degree of normalcy of the physical or mental state in which I would be left.
d) In fact, it appeared a foregone conclusion that I would be severely impaired in some capacity or another. After all, my head had been cut open; they had tampered with my brain.

And so the 'good' news was that if I lived, I would be brain-damaged.

It was no wonder that the atmosphere inside the hospital was as grey as the linoleum floors that lined its endlessly long corridors. Even in the fluorescent silver-blue glow of the overhead lights, the faces of the nurses appeared dark and serious. If anyone holding out hope for me was looking for a glimmer of optimism from the medical team, they didn't get it. Perhaps the team was trained to be like that; perhaps it was simply their way of setting expectations.

It seemed to work. Not many thought I'd make it. But everyone stayed anyway.

There was a deluge of support at the hospital. Joanna, Barney Curley and Chrissy Hills, Joanna's friend and wife of jockey Michael Hills, were among the first to arrive, having flown in on a private plane that had been organized for them. My brother Laurence and sister Geraldine had flown in from Ireland. My brother Pat had driven up from Bristol and Eamon had driven from Findon, with fellow jockey Philip Hide and Ray Watson, who was travelling head lad to Josh Gifford. My brother Michael came up from London, and my sister Maureen from Wales. The two men who used to drive me to my races, Yarmi and Jim Hogan, were both at Haydock Park and followed the ambulance to Warrington and then later from Warrington to Walton. Reputed never to really agree with each other's strong opinions, they banded together when disaster hit someone they were both tremendously – and equally – loyal to. Neither man left my bedside.

The next forty-eight to ninety-six hours would see many others coming and going, including Joanna's father,

Robert Park; her grandparents Peter and Pauline Chase; racing photographer Colin Turner; TV presenter and my neighbour in Newmarket, Derek Thompson; trainer Josh Gifford; jockeys Michael and Richard Hills. All in all, the waiting room of The Walton Centre saw jockeys, trainers, agents and other 'horse people' congregate on its worn black leather sofas in a glorious display of comradeship. The press gathered patiently outside, a constant presence. When I was in hospital, they wrote about me every day, sometimes several times a day. Most of it, like the tone inside, was bleak.

Then, just like Superman in disguise, arrived a saviour. Tommy Stack – distinctive and iconic – proceeded to pick up the shreds of despair from those grey linoleum floors and rebuild hope. According to Tommy Stack, I was going to be dancing by the weekend! The Irish trainer and former jockey was familiar with The Walton Centre, having spent four months there on account of a broken back. As such, he knew the area very well and used his familiarity to lighten the mood of several anxious people and inject some much-needed cheer and positivity into an otherwise grim environment. Tommy Stack cracked jokes, made people laugh, invited them out to the local Chinese and really got everyone's spirits up. 'This one's going to live,' he would walk around proclaiming with great authority; he even got the nurses to smile. But that's Tommy Stack for you.

In the darkest of times come the most unexpected of surprises – for me, one of them was definitely Tommy Stack. I always knew he cared about me. I just didn't realize how much.

Tuesday, 3 May 1994

At 7.30 a.m. I was seen being wheeled out of the theatre and into the Intensive Care Unit.

I was told that I came to, briefly, at the end of the operation, but I was so confused, disorientated and irrational that I was considered a risk to myself. The doctors couldn't guarantee that I would be able to breathe on my own. More importantly, they couldn't guarantee that I wasn't going to pull my tubes out, get out of bed and rejoin the race that I had been forced to abandon.

'I want Strawberry,' I had murmured in my semi-conscious daze. 'Get my horse here and let me go.' And then I'd fallen back into that deep, bottomless, chemical sleep.

'Did he want a strawberry?' they had asked each other later, amongst themselves.

And so they agreed to wait and see before they made any further decisions. I remained in the ICU where I was induced into a deeper medical coma, hooked up to a life support machine and left there until inside me, somewhere, something flickered.

Wednesday, 4 May 1994

Joanna was talking to me. I could hear the words. Her voice. Sometimes it was soft, a whisper; sometimes it was loud, way too loud – did she think I was hard of hearing?

Sometimes I understood what she was saying, sometimes I didn't. Sometimes I fell asleep while trying to understand. But when I woke, she was still there, talking to me. Just talking to me. It warmed me from the inside to know she was there. Always there.

She sounded very cheerful and happy. That was good. Joanna smiled when she was happy and she was beautiful when she smiled.

She kept referring to the things we would do when I woke up. 'When you're awake, we'll discuss our offer on Oaktree House. When you're awake, we mustn't forget to call Louise about the car. When you're awake, we will go to The Plough for that cheesecake you love.'

I wondered when I'd be awake.

Thursday, 5 May 1994

They decide to try to disconnect me from the ventilator. From the life support machine. Life Support. Those words. If you break those words down. Something supporting someone's life. Something supporting my life. Now they felt that it didn't need support any more, my life. That it was ready to go at it on its own. Or not.

I could hear Joanna's voice, fading in and out, loud and soft and loud again.

There was a man's voice, too. He was talking to Joanna. Their voices sounded distant and distorted. I could hear them, the words they were saying, but I couldn't understand them, what it all meant.

He was saying, 'What we're going to do is, we're going to take the ventilator away, and hopefully Declan is going to breathe by himself.'

Joanna was saying, 'Hopefully? How hopeful are we? What if he doesn't? Are you going to put it back on again?'

The man was saying, 'Don't worry. It's just a trial. If he can't do it, we will try again later. That's why it's called a sink-or-swim trial.'

The words echoed in my head: sink-or-swim, sink-or-swim, sink-or-swim.

Joanna sounded scared. The man sounded completely matter-of-fact.

Strange as it seems, I felt glad that I was the one lying there, waiting to be 'disconnected'. Because no matter how terrifying it was, it couldn't have been worse than standing in that room and witnessing the violence of how they tried to bring me out of my coma.

And then they did it. Just like that. In a two-step process. Step one, they disconnected my breathing tube from the ventilator and, step two, they connected it to the wall oxygen.

Suddenly, there were voices everywhere. Joanna's. And the man's. And other people's I didn't recognize.

And they were all yelling at me. 'Declan, wake up,' they were screaming. 'Declan, it's time to wake up now.' Then, 'You can do it, come on, you can do it.'

I have never been in a delivery room, but I imagine the whole environment is similar. Nurses hovering over the bed, a doctor, the screams of encouragement in those final, frenzied moments before a baby is born. Push, push. Breathe. Breathe. I suppose in a bizarre way, I *was* being born again. Giving birth, giving life. So similar, so different.

The voices grew louder. More urgent. They were shouting in unison now, like a deranged choir:

'DECLAN!'

'WAKE UP!'

'BREATHE, BREATHE!'

They were shaking me vigorously, all these people. Shaking me. Screaming.

'DECLAN!'

'WAKE UP!'

'YOU CAN DO IT!'

I opened my mouth.

There was no air.

I gulped frantically.

Nothing.

My throat seized.

My lungs burned.

'YOU CAN DO IT, DECLAN, YOU CAN DO IT!'

I couldn't do it.

I felt like I was being smothered.

Like my lungs were filling with water.

Sink-or-swim, sink-or-swim, sink-or-swim.

I wasn't swimming.

I was sinking.

Panic raged through my lifeless body. I opened my mouth to let in the air. But instead I let out a *desperate, strangled* gasp.

The voices faded to blackness. And then I felt terror. Pure and unadulterated. Until they reattached the ventilator.

Again.

Because what I didn't know at the time was that this had been the third failed attempt to revive me.

Friday, 6 May 1994

Four days.

Three attempts.

Two choices.

One decision.

After four days and three separate attempts at trying – and failing – to get me off the ventilator, the doctors considered switching off the life support machine.

At this point, the hospital issued a formal report to the press, stating that my condition had deteriorated to critical. After that, the relaying of any further information on my condition stopped.

That afternoon, Father Patsy Foley, later to become a close personal friend, was brought in to read me my last rites. With this, the press outside the hospital took their cue. They made frenzied calls to superiors: Should they wait? Should they go to print? Was it too early? But they couldn't afford to be too late! Was I dead? Or not quite yet?

The London edition of the *Racing Post* made the first bold move by preparing an unambiguous headline. It read, simply, 'DECLAN MURPHY DIES AFTER HORROR FALL'.

And that, as they say, would have been that.

If not for the strange twists and turns of a creature called Fate.

In the initial aftermath of the accident, my brothers and sisters hadn't allowed my parents to come over to England – they felt it was too dramatic a scene for them to witness until there was more clarity on my prognosis. As a matter of fact, my parents hadn't even been fully briefed on the severity of my injuries – it was a deliberate omission by my siblings to protect them from the harshness of the reality that they now all collectively faced. When my brother Pat called home to relay the news on the night of my accident, he had rehearsed two separate versions of the truth. He told my sister Geraldine how it really was: 'He's bad,' he had said, 'he's bad. I don't know if he'll make it.' But he had chosen to shield our mother from the trauma: 'He's bad,' he had said, 'he's bad. But he'll make it.'

Ninety-six hours later, when the rain had become a torrent, they had no choice but to wrench the umbrella from above their heads. Because when the doctors considered switching off the life support machine, my sister Geraldine baulked. My parents had to be told, she said to the others – after all, the decision to take one's child off life support remains solely a parent's prerogative. And so, the call was taken to ask my parents to make their way to the Intensive Care Unit of The Walton Centre where a medical professional would seek their blessings to end my life.

They say Truth is stranger than Fiction.

What happened next is the sort of stuff that would make Fiction go pale in the face.

My brother Pat called home to duly amend his story and tell my parents that they were needed at the hospital. Unbeknownst to him, my parents had already started to make plans between themselves, because hours before Pat had rung them, a reporter from a news channel had arrived at their house to get their reaction to my death, before I had even died. And so they had started – after days of denial – to finally prepare themselves for an undertaking that no parent imagines they will ever have to partake in.

Notwithstanding the dire circumstances, my father refused to fly. He had always had an irrational fear of flying and his phobia clouded all else. And so, instead of the one-hour flight from Ireland to England, my parents opted instead to come by boat – an eight-hour journey including the long drives on either side.

Ah, Providence. Oh, Destiny. Ah, Dramatic Irony.

The decision to turn off the machines was delayed by the differential of seven hours. And in that precise window of time – created inadvertently by two scared people choosing ferry over flight to cross the Irish Sea – I regained consciousness.

I woke. I breathed. I survived.

I was back.

Or so it seemed.

There is a part of a shadow that sits between complete darkness and total illumination; it is here in the penumbra, this uncertain space of partial eclipse, that I now lay. I found myself flitting between two worlds, of consciousness and unconsciousness, of half-sleep and wakefulness, swimming upwards to the surface through a black hole of oblivion. It was muddled together in the black hole, past and present, time and space, day and night. I didn't understand it, any of it.

Because when I finally woke, they spoke to me like I was a baby. Slowly. Very loudly. Enunciating every word. And I spoke back in the same manner. In the manner of a child. For the simple reason that when I woke, that is who I was – I was a child.

And Fiction gathered up her skirts and ran.

Twelve

I am eight years old.

I go to Presentation De La Salle, the Christian Brothers school in the village.

Many of the boys take lunchboxes to school, but I come home for lunch every day, because I can – my school is only about half a mile away from home.

So every lunchtime I run home.

I run at full speed.

Until I get to the last cottage on the street before the road turns a corner.

The cottage is painted a pale magnolia.

There is a drainpipe that runs along the length of the magnolia-coloured wall of this cottage.

When I run, I pace myself up to the drainpipe.

But I'm not running (in my head).

I am riding (in my head).

And as I'm riding, I'm doing racing commentaries (in my head).

My father watches the horse races on TV or listens to them on the radio, so I know the dialogue down pat.

I've actually got pretty good at it.

Of course, I'm also riding the race I'm commentating on.

And every time I get to the drainpipe, I win.

The drainpipe is my winning post.

Sometimes there is a photo finish.

And sometimes there isn't.

But I win anyway!

I do this every day, running home from school.

Every single day, without fail.

For probably two years.

For two years, I ride racehorses, running up that street, past the drainpipe.

So a couple of years later, when I go out riding, pretending to be Eamon, I know I can win because I have been winning every race I have ridden, every day, for two years.

(In my head.)

I am eight years old.

Busy, busy, busy. So busy being a kid.

One day, I am helping Mike Ryan on his farm about 2 miles outside the village, when I spot his tractor.

I've never driven a tractor before and I want to see what it feels like to be a big grown-up farmer driving a tractor.

So I climb into it . . .

And I get into the driver's seat . . .

And I start the engine . . .

And the tractor comes to life . . .

And the wheels turn . . .

I'm driving!

I can't sit on the seat because I'm not tall enough to see over the great big wheel, so I stand up.

I press down on the clutch with my foot to change gears.

I pull the accelerator with the handle on the side of the seat.

I'm driving, standing up, down the long dusty road that leads to the village, loving every minute of it.

I've just crossed the first sign for the village when Sean McCauliff, of the local guards, comes running out and stops in front of the tractor, frantically waving his arms.

I stop.

I smile.

He doesn't.

Instead, he tells me quite sternly to get off the tractor.

He points to the side of the road.

'Park it there and go home, kid,' he says.

I say, 'But I can't, it's Mike Ryan's tractor and he'll be very cross if I don't get it back to the farm.'

Sean McCauliff mops his brow in frustration and goes and fetches my father, who drives the tractor back to Mike Ryan's.

Both men glare at me.

I get sent home in disgrace.

I can't understand why.

I am eight years old.

I am a messer and a prankster.

If I ever get into trouble, I blame my sister Kathleen.

I am also a real mammy's boy. Everything I want, she tries to get it for me.

I love school now but I never did until two years ago.

I started school four years ago and I kicked up such a fuss that Mam had to get the nuns to let Kathleen start school, even though she was a full year too early.

I wouldn't go to school without Kathleen.

Mrs Kelleher was our first teacher and she was lovely.

Then, when I got older, I had to change schools and go to the Christian Brothers school with Michael and Eamon.

Again, I kicked up a fuss.

I was always trouble.

There was a handball alley at the crossroads, just past the first signpost for the village.

Every morning, on my way to school, I would go as far as this handball alley.

And I would go no further.

Michael and Eamon would be there, dragging me on . . .

But I would dig my heels in, roaring and screaming.

And then, by and by . . .

Our father would come along . . .

Straight across the road and into the shop . . .

A bar of chocolate for me.

And a clip across the ear for Michael and Eamon.

And away to school we went!

Every morning, Dad would buy me a bar of chocolate.

Every morning, I would eat it in the handball alley.

It was the only way anyone could get me to go to school.

By the time I went into 2nd Class, I had somewhat accepted the fact that I was not going to get my own way where school was concerned . . .

But the chocolate always helped!

I am eight years old.

Dad has our cows at Uncle Mikey's and on some evenings after school, we go and milk the cows by hand.

On other evenings, we play soccer on the green with our friends.

We have some great soccer matches . . .

We play uptown *v.* downtown . . .

And we are fiercely competitive . . .

So the matches can get a bit brutal at times!

Around seven o'clock, as soon as it starts to get dark, we switch to hide and seek.

By nine o'clock every night, the mothers come out, one by one, calling for us to come home.

The first kids to always get called are the Daverns – Paul and Thomas.

'Paul! Thomas! Come in!' their mother calls.

And Paul and Thomas disappear home like good little boys.

But they are always the only two to go in.

The rest of us are called about ten times before we emerge from the bushes.

Sometimes we just stay giggling in our hiding places until our mothers come and find us.

'We were playing hide and seek, Mam, we couldn't hear you!' is our excellent excuse.

Who knew it would be such fun being eight?

I am eight years old.

I have this thing with my hair.

Everyone at home complains that I spend half the day checking my hair in the mirror.

It's true, actually. The hair has to be right.

We have a fireplace in the kitchen . . .

It's an open fire . . .

And the mirror is above it . . .

Every morning, before school, we go out in the fields and ride our ponies.

Every morning, when we come back home after riding, we get washed and changed and ready for school.

And every morning, right before we leave for school, I stand on the step in front of the fire, and fix my hair in the mirror.

One such morning, I was standing in front of the mirror . . .

Fixing my hair . . .

As usual . . .

Mam was telling me to hurry up and eat my porridge.

All the others – even the girls – were already outside, fully dressed, waiting for me.

I could hear them calling my name impatiently in turns, yelling that I was making everyone late for school. So I turn around to join them, still fixing my hair as I walk away.

Suddenly, I see Mam grab a tea towel and start to beat my leg furiously.

I hadn't a clue why she was doing that until I noticed that the leg of my trousers was on fire!

So there was Mam with the tea towel against my leg, beating it and beating it and beating it while the flames came flying up from my trousers.

My brothers and sisters laughed till they cried.

And never let me forget it.

I ended up missing two weeks of school.

Mam never let me forget it.

I thought I'd sit and watch *Tom and Jerry* all day long,

until I realized that it came on only for thirty minutes in the morning and didn't start up again until five in the evening.

So I spent those two weeks lying on the sofa with my leg in a bandage, bored out of my mind.

I never let me forget it.

But it didn't stop me from fixing my hair in the mirror . . .

I am eight years old.

Kathleen and I go up every now and then to Uncle Mikey's to ride out.

He's got some great horses at his farm in Kilfrush.

He's also got these vicious Alsatians . . .

That I have a habit of teasing . . .

Only because they are always tied up with these humongous chains . . .

One Saturday morning, Kathleen and I get on our bikes and race up to Uncle Mikey's to ride out.

I end up riding faster than Kathleen, so I win the race.

I'm delighted with myself.

I even have a little private laugh.

Eventually Kathleen catches up to me, huffing and puffing.

But it turns out the laugh is on me.

Because this is the sight she sees:

My bike, on its side, lying on the ground.

Me, on my back, lying on the bike.

And Nan, the Alsatian, standing over me.

Turns out that Nan *was* tied to a humongous chain . . .

But the chain was tied to nothing . . .

Yup. Nan had me pinned and I was going nowhere.

So, my brave sister Kathleen walks up . . .

And catches the Alsatian . . .

And pulls her away from me . . .

And ties her back up safely.

I will be grateful to Kathleen the Warrior Princess for ever.

Because if I had so much as moved a muscle, that dog would have ripped me apart.

It was the last day I ever teased Nan.

I am eight years old.

Eamon is ten.

Kathleen is seven.

The three of us have such fun together riding!

Our ponies at the time are called Barney and Roger.

And what a grand time we have with these ponies!

What imaginative role-play we come up with!

What elaborate props we make!

One of the things we love to do is to go around all the houses in the estate, cutting grass.

Then we bring out the grass to the field at the back of the house . . .

There are several small ditches in the field . . .

We stack the grass alongside them, building fences.

And these fences are Aintree.

And we are the jockeys, racing fiercely over fences in the Grand National!

Sometimes we don't fancy being jump jockeys.

So we find poles and sticks and branches and whatever we can get our hands on and build show-jumping courses on the green across from home.

Then we pretend to ride the ponies show-jumping.

It's not just the three of us doing this, everybody our age living around the village gets to ride Barney and Roger as well.

We ALL have turns.

So poor Barney and Roger end up going around a LOT of times.

Then, when the ponies get tired, we let them off and jump ourselves!

So now we are both the horses *and* the jockeys.

We spend hours every day doing this . . .

It's mad fun!

I am eight years old.

Liam Jones has a shop where most of the village does its shopping.

It's also where the phone is.

This is the phone my brother Pat rings to speak to my mother.

Pat is out on the Curragh now, busy being a jockey.

We don't have a phone in the house at this time, no one does; colour TV is just coming in . . . exciting times, these.

So Pat calls Liam Jones's every Friday evening at 5 p.m., like clockwork.

Liam Jones sprints across the street to get my mother.

My mother sprints back with him.

OK, maybe 'sprint' is an exaggeration.

Liam Jones also has a farm about a mile outside the village.

This is where he sources most of the supplies for his store.

In my summer holidays I help him out on the farm.

I'm a very helpful little boy.

One time, Liam Jones wants a hand stacking the hay in the donkey cart to bring back into the village.

So, one afternoon, I run up to Liam Jones's.

He tells me to take the donkey and cart up to the field and says he will meet me when he has finished up at the shop.

So up I go and I wait for him to come.

I sit patiently.

And

I sit patiently.

And

I sit patiently.

And then I run out of patience.

So I start loading the hay myself.

There are about a hundred bales of hay, but it doesn't seem daunting. It's far better than waiting.

So, one bale at a time . . .

I load . . .

And

I load . . .

And

I load . . .

And before I know it, it's done.

But there's still no sign of Liam Jones.

So I scramble up the fully stacked cart.

And perch myself above the hay.

And ride the donkey cart back to the village just as Liam Jones is getting ready to come and meet me.

I must have made for a funny sight . . .

Because he lets out a great big laugh.

'Jesus, did you load that all by yourself?' he says. 'You great boy, Dec, you!'

Then he runs back to the store and comes out with his camera.

Still laughing, he takes my picture.

Here it is.

Me, as an eight-year-old kid, no more than four feet tall, grinning cheekily as I sit on the very top of a donkey cart, piled six feet high with hay that I've stacked all by myself.

Did all of this happen only *four* years ago?

Loss

The first marker of success in brain surgery is when the patient wakes up.

The second marker is much more imprecise. This pesky little nugget could take months, years, for ever, never.

When I came out of my coma I was mentally twelve years old.

There were three men standing above me. They asked me all these questions, almost like they were quizzing me. I was tempted to laugh; the questions were so easy. I answered them all instantly.

What country do you live in? *Ireland.*

Who is the Prime Minister? *Jack Lynch.*

What horse were you riding? *Strawberry.*

Silence.

How old are you? *Twelve.*

The men in white coats looked at each other with grave faces. Then, without another word, all three turned around and left.

See no evil. Hear no evil. Speak no evil.

★

After the trio of neurologists left, my family came in briefly – I had been moved from the ICU and into a private room at this time. Kathleen held my hand, and Pat and Eamon and Michael and Geraldine and Laurence and Maureen stood at the foot of the bed. Joanna sat down on the bed next to me. My parents, almost too scared to look at me too closely, stood apprehensively by the door. Everybody was speaking all at once, trying so hard to make things appear normal. But of course I knew nothing was normal. They were only there for a few minutes before they were ushered out by a nurse.

When everybody left, a man dressed in a tweed blazer and navy-blue trousers entered my room and closed the door behind him. Then, leaning against the door, hands casually in his jacket pockets, he introduced himself to me as Professor John Miles, my brain surgeon. I recognized his voice instantly, I could never forget that voice – it belonged to the man speaking to Joanna when they were attempting to take me off the ventilator.

Then, he perched himself on the edge of my bed, body sideways, leaning forward – he was so short, I noticed that his feet didn't quite touch the floor – and uttered the words I will always remember. He spoke in slow, measured tones – not unkindly, but if you are ever under the illusion that a doctor breaks bad news to you with an ounce of sentimentality, think again. Professor Miles was as matter-of-fact as they come. Not that I blame him; there is rarely a gentle way to deliver a knockout blow.

'Declan,' he said, 'you've had a very bad accident. You've had brain surgery. There are going to be some consequences to this – to what extent, we don't know. Initially, there won't be any sensation in your arms and legs; you'll be able to move them, but you won't feel much. You won't be able to stand or walk. This is because you are paralysed – to what extent we don't know. One of your blood clots . . .'

All the time, while he was speaking, I was watching him, half listening to his words but mostly thinking that this man looked nothing like I expected a brain surgeon to look. I'm not exactly sure what I expected a brain surgeon to look like, but Professor Miles certainly didn't fit my bill. He was a short, neat man with a receding hairline and smart eyes. I liked his eyes, they were blue with a twinkle about them, an intelligence. He was droning on, this man with the smart eyes who had just cut open my skull, giving me more reasons to celebrate the fact that I was alive.

'. . . was a hair's breadth away from your optic nerve and we are quite likely to have damaged it. The optic nerve doesn't like to recover, so you might have some blindness in your right eye – to what extent, we don't know. Other nerves, many of your cranial nerves, were also disturbed. You might have complications develop with your speech and your hearing – to what extent, we don't know. And most importantly, it appears that you have suffered some memory loss. There are lost years. We don't know how many. But you're in a safe place now. We will look after you here.'

At this point, he hesitated. Just ever so slightly, before looking me straight in the eye – and finally, conclusively, revealing his hand. 'The problem is we don't know when any of this will come back, how much will come back, or if it will come back at all. It could take years to find out.'

He stopped then, waiting for a reaction from me.

It didn't come. His words did not even register. He could as well have been talking about how he liked his steak cooked. Or his shirts ironed. It was all the same to me. I listened with polite disinterest.

You probably find this odd. Very odd. You probably would expect his words to have had a devastating effect on me. As might happen when everything you've ever known about yourself is swept away in a spinning vortex. Or when you realize that the specks of dust flying around you are actually the crumbled bits of your broken dreams. Or when your brain surgeon tells you that you've almost certainly lost every hope of living a normal life.

So yes, you'd be completely right to think that his words might have evoked *some* kind of emotion in me. Maybe the usual suspects: sadness or shock or fear or anger or anxiety. One or some or a heady cocktail of the lot?

They didn't.

A bit of nervousness, maybe? A tear or two?

Sorry, no.

I had absolutely no reaction. While 'Professor John Miles, Brain Surgeon' was telling me that I had lost years of my life, that I might be paralysed and more than likely be blind in one eye, I was thinking the whole time about how this brain surgeon didn't look like a brain surgeon at all.

I would later reflect upon my reaction with a mixture of surprise and disbelief. Not because it was so abnormal, but because it was so ridiculously *normal.*

I was often jokingly referred to as 'Iceman Murphy' by my riding colleagues for staying cool under pressure, for my resilience in the moments that mattered. Never weaken when you need to strengthen, I had taught myself. And I had tried to follow this in every setback that life had thrown my way.

But this time was different. These were murky waters riddled with treacherous unknowns. A formidable challenge, even for the most unflinching of us. And yet, I had treated it in the same way as I would any other mishap. What would amaze me – years later, when I had time to reflect upon this – was how one never really wavers from that initial gut instinct, despite the magnitude of the circumstances. It was subconscious, of course it was subconscious, but it didn't matter that I had suffered one of the most horrific accidents in racing history. It didn't matter that I was a twenty-eight-year-old man, who had woken up as a twelve-year-old child. My reaction was exactly the same as it would have been if I was in a race thinking my horse would win, and then realizing, suddenly, that it wouldn't.

I didn't break down. I didn't show emotion. I stayed strong.

It didn't matter the degree of crisis. A crisis was a crisis. And I *had* to stay strong.

Maybe John Miles had made my decision for me. He had laid out the specifics of my situation in excruciating detail. There was nothing left to be done or discussed. And so it seemed, pragmatically, that my reaction was almost irrelevant. What did it matter how I reacted, or whether I reacted at all? It wouldn't change a thing. I had been given the irrefutable facts. And the facts of my condition rendered my reaction meaningless.

It wasn't that I thought he was wrong.

In fact, I *knew* he was right.

He was dead right.

My memory was wiped out.

Clean slate on a Monday morning.

I couldn't remember much. Whatever I could remember seemed so recent, and yet so far away. It was all terribly mixed up inside my head; a bizarre amalgam of disparate events that made little sense, even to me. Somehow, in all of them, we – myself, my brothers and sisters, my friends – were all children.

I remembered my brother Michael teaching us to drive the old Anglia Estate around the field. I remembered doing the hay for Miley McElligott and Liam Jones in my summer holidays, bringing it in by donkey and cart. After that, I started doing the hay for Mike Chairman Ryan on a tractor. This was all about four years ago, so I must have been eight years old at the time.

I also had some memories of riding. Everyone said I rode well. The riding memories were much more recent – vivid and fresh in my mind. I had just won the champion jockey's title, pony racing. I remembered riding many winners for Tommy Walker from Newcastle West. He had a pony with two names. Her name was Daisy, but she was also called Bluebell. Eamon and I would tease her, each standing at one of the two far corners of the stables, one of us shouting 'Daisy!', the other shouting 'Bluebell!', and watching her turn her pretty head from one side to the other in response to her two names! Another pony I remembered riding a lot of winners on was String of Pearls for Christy Doherty, also from Newcastle West. Paddy Shaughnessy in Galway gave me a lot of winners to take the title. I remembered him, too.

My recollections of riding couldn't have been from further back than a few days. Or perhaps a few weeks. But not much longer than that. I was just starting to get too heavy for the ponies. I thought Kathleen might win the title from me. But it didn't make sense because she had just been here, by my bed, a woman. And so were my other brothers and sisters, standing here in this room in front of me, a few minutes ago. And they weren't children either. And neither was I. We were all fully grown men and women. How could it be?

That was me, mentally.

Physically, I woke up not feeling. Not feeling anything. My arms, my legs, they were numb. Sometimes I could move them, sometimes I couldn't, but I couldn't feel them at all. There was my head and my face. Then my torso. And then nothing. I found at times that I could reach out and grab something, while at other times I couldn't. I couldn't control when I'd succeed and when I'd fail but, even when I did succeed, it didn't mean much because I could never feel that something. Hot, cold, pain, nothing. The messages from my brain refused to transmit to my limbs. I couldn't stand, I couldn't walk, but finally I could breathe, and I suppose this is when you shake your head and say, 'What a lucky, lucky man.'

So, of course, Professor Miles was right. And yet I ignored him. From deep within my subconscious, I blocked him out. And with that, in one fell swoop, I blocked out *everything* that *anyone* thought was wrong with me.

It wasn't denial.

It wasn't madness.

It was self-preservation.

A few minutes after Professor Miles left the room,

Joanna and my friends and family came in again to see me. They must have been given instructions not to stay long this time around either, because they had barely spent a few minutes with me when they told me that I needed to rest and they needed to leave. I made a sudden motion then, as if to get off the bed, but of course I couldn't – the message from my brain couldn't get to my feet. My sister Geraldine looked at me in horror and said, 'Declan, what on earth are you doing?'

I said, 'Hang on, I'm coming with you.'

There I was, wired up, four days recovered from twelve fractures to my skull, and I believed – truly believed – that I was ready to get up and go. There was this trauma I had suffered, but it was as if everybody else was suffering from the trauma, except me. There was something that had happened to me, but it was as if everybody else was seeing what had happened to me, except me. I was fully protected. Some unexplained, inner safety mechanism shielded me from it. I refused to see it. I refused to suffer it.

But I felt it. The burden of it. Of this life after death.

Because you cannot choose not to feel.

You can feel, but choose not to acknowledge.

You can acknowledge, but choose not to act.

But you cannot choose not to feel.

So I felt it. Every mystifying moment of it.

Night and day were the same to me. I slept a lot. Often, too often, I found myself drifting, floating, giving in to the fictitious, pretend sleep of the painkillers. I dreaded this sleep. I had no control of my mind when the drugs took over. I would wake, screaming from the nightmares that tormented me.

But when I woke, I woke happy.

A happy child in my home in Ireland.

Then I would fall asleep again.

And then I would wake.

Happy child. Home. Ireland.

Sleep. Wake. Sleep.

Sometimes, I would fight to stay awake, fight with every last inch of resolve. But mostly, I'd lose. Mostly, I'd give in to whatever the sleep, whatever my drug-induced mind, chose to bring my way.

It was completely whimsical, my mind. It was morbid, it was cheerful; it was out of control. It had a life of its own. It taunted me constantly. I had no concept of what was real and what was imagination; what was fact and what was fantasy. I didn't know when I was awake and when I was asleep. I didn't know if I was having dreams or nightmares.

And I didn't know if the dream was my childhood or if the nightmare was what I was living through.

Kaleidoscope

It was a dream. It was real. It was a dream.

I was twelve and I was ecstatic.

Some time ago, I had been given Roger. Bruce gave him to me, the manager of Kilfrush Stud; gave him to my father, actually. Roger was French. He had been brought over to Ireland by the owners of Kilfrush Stud but he proved such a temperamental pony that neither their daughter nor Bruce's son Claude wanted him. So Bruce asked my father if one of us kids might like to try riding him.

What first struck me about Roger was his long, uncontrollably wild mane. It was thick and knotted, and looked as if it couldn't ever be tidied up. Many years later, I would go to Hermosa Beach in California while riding track-work for Charlie Whittingham and, in my muddled-up imagination, I pictured the human equivalent of Roger to be a typical LA 'surfer dude' with long blond hair and a free spirit. Roger's personality mirrored his looks – I loved his eccentricities and his extravagances. And so, with much fanfare, I became the owner of a beautiful chestnut Shetland called Roger.

Eamon, by then, already had his own pony, gifted to him by way of Pat Hogan. The pony had been left with Pat Hogan for safekeeping by his owner – a travelling salesman named Barney Sheehan – when his trailer broke down by Pat Hogan's yard. The trailer was duly fixed but Barney Sheehan never came back to collect his pony. Thoughtful and considerate as ever, my brother named the pony after his negligent benefactor. Barney was a dark bay pony with a hog mane – obedient, kind, easy to ride, everything you would want in an animal, really.

I discovered quickly that Roger couldn't have been more different; he truly was the most difficult character you could imagine – you couldn't even get a bit in his mouth. The moment you walked into the paddock, he would run everywhere, do everything, so as not to let you catch him. And then, when you finally got a hold of him and started to ride him, he would gallop about with a mind of his own, spontaneously stopping, about-turning out of nowhere, going in the opposite direction to where you wanted. I realized that to ride Roger, you had to have your wits about you all the time.

One day, with the fearless optimism of children, Eamon and I decided to tie the two ponies together by their head collars and ride them around the field – Eamon on Barney and me on Roger. We thought that because they were tied together, Roger would be forced to do what Barney was doing, and because Barney would obey our command, the obstinate *cheval* would have no choice but to follow. So we pointed them both at a stone wall that we intended them to jump over. Lo and behold, when we got to the wall, Roger decided to duck instead of jump, and when he did, he pulled Barney straight down with him. Eamon and I went flying in opposite directions straight off their backs, landing on the ground. It was such a stupidly dangerous thing to do, but we lay on the soft green grass like that for a long time, just laughing our heads off. It would always remain one of the happiest days of my childhood, at one of the happiest times of my life.

It was a dream.

It was a nightmare. It was real. It was a nightmare.

I was twelve and I was hungry.

When I awoke, it was dark. I sat up in bed and rubbed my eyes. I was in our home in Bank Place in the pale-blue-coloured bedroom I shared with my brothers Michael and Eamon. A stray bar of moonlight crept through where the curtains, cut ever so slightly too narrow, wouldn't quite draw shut. I could make out the faint shapes of Eamon and Michael in their beds, fast asleep, their bodies rising softly with the rhythm of their breath. I was hungry; I decided I wanted a hot, buttered piece of toast.

As gingerly as I could, so as not to wake my sleeping brothers, I crept out of bed and tiptoed across the floor. Michael stirred, and then in a sleepy voice, whispered, 'Declan, is that you? Can you get me a biscuit?' before falling straight back to sleep again. I stifled a laugh. Michael was sixteen at the time, four years older than me, and one of the most gifted storytellers you could imagine. Every night, I'd bribe him with biscuits to tell me stories – he'd make them up in his head and come up with new and fantastic ones every time! I wondered what he'd conjure up tonight when I brought him up his biscuits. Smiling to myself, I shut the bedroom door behind me softly. Then, arms outstretched, I groped my way along the corridor wall to the stairway that would take me downstairs to the kitchen. I dared not turn the hallway light on for fear of waking my parents. In any case, my eyes had adjusted to the darkness by now and I could see the landing of the stairs before me.

We had a long, rectangular mirror on the wall facing the stairway, which came into sight about halfway down. It was at about this point that I sleepily rubbed my eyes and, almost by force of habit, looked up at the mirror.

My eyes opened wide in horror.

For the person looking back at me from the mirror wasn't me.

Only it *was* me.

But it wasn't the face of a twelve-year-old child.

It was the face of a man.

It was a man and he looked like me.

It was me and I looked like the man.

Then, suddenly, around my face – the man's face, the face in the mirror – others appeared. The faces were shapes, an amalgam of ovals and circles, white and distended, with skull-like eye sockets, no noses and round black holes where there should have been mouths. There were several of them in the mirror, floating around. Then, all at once, they seemed to see me, their deep sunken eyes converging in unison on my face in the mirror.

I tried to scream, but my voice had no sound.

I tried to run, but my legs wouldn't move.

I could only watch, frozen in terror, as the faces started shaking before me, twisted and grotesque, up and down and up and down. The eye sockets became slits, each lipless mouth gaped wide in a hideous O.

They were laughing. Laughing at me. Silently, soundlessly.

Then, the sound came. But it came from me. From the man in the mirror. He opened his mouth. And he screamed and he screamed and he screamed.

It was the shrill shriek of a child.

I sat up in my hospital bed, still screaming uncontrollably, until the nurses came to sedate me.

It was real.

Umbra

Time became an abstraction.

The second hand stalled.

But somehow the minutes rolled into hours, and the hours into days.

Nothing made sense. The hallucinations continued to haunt me. Even when there was daylight, even when I was awake, it didn't stop, this ever-moving, ever-changing kaleidoscope of images that danced inside my head. I was dizzy from the confusion of it all.

What was real? What was not? And did it even matter?

I tried to keep myself busy to fill the days, to make the time go faster. *Until what?* I would often wonder. Still, I insisted that my blinds were never drawn, not even at night. And I waited for sunrise every morning, staring out of the hospital window, counting down the minutes, thinking, when the sun finally came up, that somehow I had survived to see a new day.

Some days, I would occupy myself by going through the sackloads of mail in my name – the hospital had been deluged with cards and letters for me, and a whole army of people were wading through it all. Joanna and her father took the lead on this. They would keep aside some of the get-well cards and the 'we are praying for you' letters for me to read myself; the rest they would read aloud to me. But they made sure I was aware of every single missive sent to me by well-wishers; friends and fans alike. I was told later that Robert Park had personally responded on my behalf to everyone who had sent something in. It took him weeks.

Did I mean so much to so many?

One of the most memorable letters I received was from a lady in the Midlands, who wrote to say that she was not a racing fan. But, she continued, she had put her six-year-old son to bed one night and then heard voices in his bedroom. She went in to see who he was speaking to, and apparently he told his mum that he was saying a prayer for Declan Murphy to get better. I had no memory of ever being the man that her little boy was praying for, but it was things like this that gladdened my heart, that made those long, dark days drift by.

Sometimes, I asked different people for their version of what had happened to me at Haydock Park. Since (thankfully?) I remembered none of it, it was like being told a story, each version with its own twist, coloured by the perspective of the storyteller, but all equally fascinating to hear. I listened, amazed, in a state of disbelief. Like I was being told about someone else. Like I was privy to a bit of juicy gossip. *Really*? I wanted to ask. *That* actually happened to him? To him? To me . . .

It seemed that I had shaken the world slightly by not dying. It is an unusual thing to think about oneself; a ridicule of realism, strangely transcendental. But it did appear that everyone had given up hope so long ago, that my dying wouldn't have sent as many shockwaves as my *not* dying. The *Racing Post* had to pull from publication the headline announcing my death. Instead, the now infamously premature front page was framed and hanging behind the desk of the editor, Alan Byrne, in his Canary Wharf office.

My manager, Marten Julian, had been telephoned and subsequently interviewed by Sky News at 6 p.m. on the day of the accident, all within three hours of my fall – *before the body was cold* – and told to assume that I was dead. More specifically, and much to his horror, he was advised by the reporter that, when referring to me in the interview, it would be prudent to speak of me in the past tense. Marten claimed that the pressure had been 'almost unbearable'.

Separately, Barney Curley told me that just going by the accounts of *Racing Post* journalist Ray Gilpin, who spent every waking moment at the hospital monitoring my progress, he had stacked my odds of living at a 6 or 7/1 chance in a two-horse race. Barney Curley rarely ever lost a gamble in his life, but luckily for me, I suppose, he must have lost his Midas touch on that one occasion. Racing photographer Colin Turner, standing closest to the scene of the fall, had witnessed the horse galloping over my head and said that he had never heard a sound like it in all his professional career, that the 'sickening thud of hoof hitting helmet' would regrettably stay with him for ever. Indeed, Eamon told me that Arcot's owner, American businessman Jim Chromiak, rang him up minutes after the fall and wasted no words. 'He's died,' he told a shell-shocked

Eamon on the other end of the phone line. Apparently, I had managed to prove him wrong as well.

I thought it an extraordinary coincidence that the horse that had delivered the final blow to my skull was being ridden by Charlie Swan. I didn't have any recollection of this, but Charlie told Eamon that we had been standing together in one corner of the weighing room – just he and I – minutes before the start of the race, discussing Ayrton Senna's untimely death and pondering our own mortality. And then, by a peculiar twist of fate, *he* happened to be riding the horse that had almost caused me to suffer the same fate as Senna. Honestly, could you make this stuff up if you tried? Of all the possible riders in the race, did it have to be Charlie?

But it didn't just stop there. After Cockney Lad had galloped over me, too quickly for Charlie to react despite his best efforts, Charlie had apparently been in such a state of shock that he was going to pull up halfway up the run-in and go back to see if I was alive. Instead, he rode on, ashen-faced, drenched in a cold sweat, not daring to look over his shoulder. You see, what happened couldn't have been prevented, but it was emotional. Charlie Swan was much more than just a colleague, Charlie had been my childhood friend – we would go pony racing together as ten-year-olds, two Irish kids with long blond locks, ready to conquer Rome. After my accident, this nine-time Irish champion jockey told Eamon that if I didn't live, he would never ride again.

The irony of it all boggled the mind.

Of course it also boggled the mind that I was thinking about it lying in a hospital bed that I couldn't physically get out of even if I tried, because my legs didn't work. Well, at least Charlie would ride again. The silver lining. What would we do without the silver lining?

Often as I sat alone in my room, I pondered all the 'what if?' scenarios. I did this to keep my mind busy, to pass the time. There were so many of them, it became a little game. I would recount them in my head, over and over again.

What if my mind hadn't been clouded by Senna's death?

What if I hadn't been riding the one horse that had tried to kill me the last time I'd ridden him?

What if I hadn't already been knocked unconscious before hitting the ground?

What if I'd fallen close enough to Arcot for Cockney Lad to clear us both in a single stride?

What if I'd fallen far enough from Arcot for Cockney Lad to clear us both in two separate strides?

Would the outcome have changed if any one of these had played out differently? Of course, I would never know, but I wondered all the same.

Sometimes, I thought about the theatrics of it all. I was a child in my head and I felt a childlike thrill at how melodramatic it had been – the police escort, the sirens, the jockey who'd fallen off a horse. I've always loved drama like this – you see it in the movies and things, and here I had been living it. It seemed such a pity I was asleep through it all.

Other times, it made me philosophical. Lots of people told me that if my circumstances hadn't been as high-profile as they were, there wouldn't have been a police escort, Professor John Miles most likely wouldn't have been called in on his day off, things wouldn't have been done with the urgency and the efficiency that they had been. Did my perceived fame then save my life? Should I be thankful for it? Should I feel guilty about it? If I was an anonymous man, if I wasn't Declan Murphy, would they have let me die? Why was my life any more or any less precious than that of the man next to me? How does one measure the value of a life?

I thought about this, about moral absolutes in a relative world. Freedom and justice and life. It was odd to think about these things that seemed so grown-up and yet I thought about them – I couldn't help myself. I was a child with adult thoughts. I found myself fighting this paradox constantly.

One day, Joanna brought me my obituary to read. It was a genuine, highly eloquent, fully completed obituary of me, Declan Joseph Murphy, for the *Racing Post*, penned by talented journalist and my good friend Paul Haigh. It had been written and edited, ready to circulate, stopped only in the nick of time; a comprehensive and emotional fact file of my life (and death), career highlights, notable wins, and most interestingly, how I would be remembered, as a jockey, as a person, as a friend . . . I read it lying on my hospital bed, over a cup of heavily diluted tea and half a mushed-up stem ginger biscuit, and I was so astonished by it all, I nearly toppled off the bed, wires and everything. It was amazing the things people said about me.

I mean, you may not realize this, but you're fucking *amazing* when you die.

But, here I was – not quite alive and kicking, but alive nonetheless. And I didn't really know what to do with myself. On some days it seemed an endless journey home.

Slowly the years came back, the consciousness, the

memories – as if they were being drip-fed through one of the many tubes hanging from my body. But ultimately, once everything that was due to be returned had been returned, I had to accept that some of what was mine had been snatched from me.

When I had woken up from my accident, almost two-thirds of my life were missing. Within days, I got back all but seven years. More gradually after that, I got back another three.

Everything came back chronologically but disordered. At this point, I could recall distinct snippets of my life, of my past, but I also knew how wrong they were, how so much of it just couldn't be. They were jumbled together in my head, memories, images, from different times, different places; some possible, others improbable, all mixed up into a riddle I still couldn't solve.

I clearly remembered riding Strawberry, our family's workhorse. I had just ridden him to deliver milk to the creamery. I had left him there outside, and when I came back, he had disappeared. Was this real? I couldn't be sure. I was five. Then I found him again on the Curragh when I was riding gallops on a different horse. He was just standing there, alone on the flat, open plain, watching me ride past. Was this real? It couldn't have been. I was eighteen. And then I was with Joanna. We were at Browns in Cambridge. I was all dressed up in an ironed shirt. There was a piano player. Joanna was laughing. We seemed so happy. Was this real? This seemed more probable. I was twenty-three. And then, my last memory of all – the Sunday of the Prix de l'Arc de Triomphe in October 1989, in the garden of Joanna's grandparents, Peter and Pauline Chase.

It was the first time I had picked up a golf club in my life and Peter Chase was teaching me how to swing it. Was this real? It seemed so vivid. I was not quite twenty-four. Then it all went blank. And I could remember no more.

So I waited. I waited for what was rightfully mine. The missing four years. My missing four years. My precious, *precious* four years. The four years leading up to the accident. The four years that defined my racing career. The four years that culminated in everything I had worked towards. The four years in which I had found success, fame, love.

So I waited. And I waited. And I am waiting still.

For they never came back.

I regained only part of my memory; the most recent four years of my life were gone for ever.

And interlaced with this, so tight it could not be wrenched apart, was my sense of identity.

There is a time that comes when science confounds itself.

Life goes backwards, memories disappear.

And darkness moves faster than light.

This was that time.

This was my umbra – the darkest part of a shadow, the darkest part of a life, a total eclipse.

The umbra of the Earth is 1.4 million kilometres long.

Mine was humbling in comparison.

But it was equally dark.

Because no matter how strong I was, how hard I tried, how much in control I felt, there was something I simply could not do – I could not remember a part of myself.

Hubris

How does a man look to the future when he cannot remember his past?

I had lost four years of my life.

Try as I did not to let this affect me, not to consciously think about it, there it was like a relentless rerun of a horror film.

Inside my head.

Turned up to full volume.

Dominating my existence.

I couldn't switch it off.

I had lost four years of my life.

And now? What lay ahead and did I dare to find out?

I was one of the lucky ones – I didn't die; I lived. I lived through my accident, but this – living the *consequences* of my accident – seemed a challenge more crippling than death itself, because I couldn't understand who I was before it, and who I was supposed to be now. And this utter lack of comprehension over my own identity created mayhem inside my head. Both that and *unfathomable* pain.

The physical pain, I could accept. In fact, when I felt it – when the drugs allowed me to feel it – I welcomed it.

Pain meant sensation and sensation meant hope. But the emotional pain was exhausting in its confusion.

Ten days after brain surgery, I made a decision of unimaginable boldness, foolhardiness and unprecedented risk. I decided I didn't want to stay in the hospital, I didn't want the care, I didn't want the medication. All of it – the environment, the carers, the drugs – was making me feel unwell.

There was no reason for this that can be explained away by logic. It was instinct, pure and primal. And I trusted in my instinct implicitly. Intuition had always been the wisest force in my life, the inner voice that spoke without words. And when it spoke, I listened. Sometimes it whispered, sometimes it screamed. But I always listened.

Now, my instinct was a roaring dragon.

Pacing. Raging. Breathing fire.

A caged animal.

Suffocating in a prison of incapacity.

No, I had to leave.

Intuition is not meant to be institutionalized.

When Professor Miles was updating me on my condition, I never accepted that there was anything wrong with me, because something inside told me that I was going to put it right. I was a sportsman; my cells were programmed to think like this – to never accept adversity, to never weaken when I needed to strengthen. In my head, I saw it all as a temporary state and I saw within myself the power to conquer it. I knew that I may not be able to conquer it right now, I may not be able to conquer it today, I may not be able to conquer it tomorrow, but I honestly, truly believed that I would conquer it. I believed this. I believed it in my head. And my head had never let me down.

Throughout my career as a jockey, I had been set apart from my colleagues, not only for my riding skill, but also, notably, for my intelligence. And I had worked the two in unison. Using a combination of gut feeling and judgement, I had developed a strategic, perfectly instinctive approach to race-riding. Before I even got on a horse, I had a game plan, measured and calculated. I would analyse everything, pre-race and post-race, strive to correct past mistakes, strive to surpass past expectations. And then, when I got on the horse, just while cantering to the start, I would consider my options, judge my pace, determine my tactics – I wasn't riding for show; I was riding to win. And so every stride was thought through with careful deliberation, split-second decisions were taken with split-second precision. I didn't just go out and ride. I used my brain.

Now, in the harshly revealing fluorescent glow of hospital light, I found it terrifying that I might not just lose my riding skill but I stood in danger of losing my brain. The grand irony of it all was that the horse's hoof had not struck my neck or my shoulder or my arm or my leg – it had struck my brain. Wherein lay the intelligence that defined me. This frightened me far more than losing the ability to ride. This threatened the very essence of my being. I looked down at my hands – I had lost total sensation in them – and I realized that this could take *everything* away from me.

We die. We all die. But after you stare into the heart of death and come out, somehow, on the other side, you understand what it means to live.

So I held my head, the one that had been galloped upon,

in my hands, the ones without sensation. And if I felt despair, I didn't show it. Because I knew no matter how damaged it was, my head would never let me down.

It didn't matter what the doctors said. It didn't matter what the experts believed. My intuition was always right. The eclipse **would** fade. The umbra **would** lift. I **would** recover.

Pardon my hubris.

Strength in Numbers

Documented Falls in the British National Hunt Season, 1993/94

My record: 10 falls in 385 rides

Or: 1 fall every 38.5 rides

Jump jockeys average record: 1 fall every 9 rides

Stated another way: To suffer 10 falls, the average jockey would ride only 90 races versus my 385 races.

Stated another way: In close to 400 rides, I fell less than the average jockey would fall in 100 rides.

Stated another way: In the 385 rides that I fell 10 times, the average jockey would have fallen 43 times.

My odds of falling: < 3%

My odds of living: < 50%

Lies, damned lies, and statistics.

Joanna

There is no book without love. Because there is no story without love. My story had love, and her name was Joanna.

I first set eyes on Joanna Park in the summer of 1989. It was towards the end of August and I had just come back from California, having spent six months with Charlie Whittingham, while also doing classes at UCLA. I was still undecided about my future, with racing on one side and a law degree on the other, and me in the middle, stuck at a crossroads. And then I met a girl.

My first encounter with Joanna was, quite aptly, on horseback. Barney Curley's stables in Newmarket were down the road from Charlie King's livery yard, and often when riding out for Barney, we would spot Charlie's string, riding across the village of Exning.

This is how Joanna and I first met – with our eyes. So stunned was I by the vision in front of me that I couldn't even bring myself to say hello. Instead, I sat stupidly on my horse, while she rode past me on the other side and disappeared just as suddenly as she had appeared.

But what's for you, will not pass you by. I believe this.

And meeting Joanna was for me.

A few days later, I saw her again. Porky (Michael O'Rourke) and I were outside by the entrance to the stables and I saw Joanna drive past in a car. 'Quick,' I said to Porky, 'that's her.' So we got in the car to follow a girl. I'm fast on horseback. I'm faster in a car. But I would quickly learn that Joanna was more than a match for me. To my frustration, I lost her again.

So I went next door to the livery yard and asked Charlie King who she was. He said, 'Oh, she's just a schoolgirl. She lives there in that big house across the road.' And before I could think very much about what I was doing, I went and knocked on the door of 'that big house across the road'. And there I met her father, a man who was to become a friend and an ally from that day on. Joanna, he told me, lives with her mother, and her mother lived on the other side of Cambridge. But he said he would give her the message. 'This strange Irish boy in a great big car came looking for you,' Robert Park would later tell his daughter Joanna.

And so it all began.

A while later, Robert Park invited me to his house to meet Joanna. I fell in love with her right away. She was beautiful, she was kind; there was a softness to her that touched me deeply. And right here, in the middle of my love story, something unexpected happened. Something I look back on with a warmth that fills my heart, something that makes me proud to be me. And it is this: I presented myself to Robert Park – reeking of aftershave and dripping with confidence – with the sole purpose of trying to find Joanna. But in doing that, I gained in him a friend for life.

It was the same with Joanna's grandfather. I first met

Joanna's grandparents, Peter and Pauline Chase, on Arc Day 1989 at their house in Woodbridge. It was on the Sunday of the Arc, the first Sunday in October and a memorable day in racing history – Carroll House, trained by Michael Jarvis and ridden by Mick Kinane, had won the Prix de l'Arc de Triomphe over 2,400 metres at Longchamp.

I connected with Peter Chase straight away. There were fifty years between us but they were of no relevance to me; we could sit together and talk for hours. I couldn't get enough of him – his wisdom, his humour, his intelligence. He used to write short stories, and he enjoyed reading them aloud to me – he had the most perfect command of the English language, this wonderful ability to craft beautiful-sounding phrases, and I could have sat with him all day long just listening to him read. I was so moved by the way he spoke – the emotion in his words, the meaning behind them, his diction, his voice. I still remember his voice, the quality of his voice. When I close my eyes, I still hear it.

The warmth with which I was received by Peter Chase until he died, the warmth with which I am received by Robert Park, even to this day, tells me that somewhere, in some way, I had to have been a good person. And this means everything to me.

Meanwhile, the lay of the land was such that Joanna's father had become a great friend, Joanna's grandfather had become my *best* friend; Joanna, however, was still not my girlfriend. Sadly for me, she didn't fall for my charms immediately. But happily for me, her father did. And that certainly helped. Of course with *that* kind of momentum on my side – for Robert Park is as enthusiastic, persuasive and charismatic a man as you will find – it was inevitable that she would succumb eventually. And after a month of asking her out every day, eventually, she did.

Meeting Joanna simplified the decisions in my life. I was twenty-three, Joanna was eighteen and on her way to the London College of Business, while at the same time starting to establish herself as a professional model. I decided I didn't want to go back to California. I was ready to give it all up; after all, I had met a girl. Instead, I decided to follow her to pursue my education. At exactly this time, I began to capture the attention of Josh Gifford; not long thereafter, I was offered the much-coveted job as his stable jockey. And once again, I found myself at an all too familiar place – the crossroads. My education on the one hand; one of the best jobs in the country on the other. I took the job with Josh Gifford.

I continued to see Joanna while I rode for Josh Gifford. She transformed my entire view of the world. I was a boy when I met her; Joanna made me a man. I remember that period in my life like it was yesterday. Starlit skies and warm summer nights. My career advancing with every ride, beautiful Joanna by my side. We were the perfect couple in a perfect world – glamorous, successful, exciting. Much like a painting by an Old Master, our life had a distinct and magical allure – even a speck of imperfection was viewed by the beholder as an intentional brushstroke. I couldn't have asked for more. Everything you could want from life was mine.

Then I had my fall. And the course of two lives was irretrievably altered.

Yes, mine . . .

But more importantly . . .

Hers . . .

At the time of Declan's accident, we had been together for five years. We were deeply in love, practically married. He had come out of the mist, this Irish man with the intense blue eyes, ridden up to me on a horse, quite literally, and swept me off my feet. I loved our life together; it was a storybook romance. Except I was living it.

2 May 1994 was like a knife that cut my perfect life into two clean pieces. The Joanna Park that went to sleep on 1 May died that night. The next day, at approximately 2 p.m., a new Joanna Park was born, changed for evermore.

On the morning of the Haydock Park races, I awoke with an uneasy feeling. Declan and I had discussed Senna's death the previous night and I was full of misgiving that morning as I saw him off to the races. Being a jockey's wife or girlfriend, you are constantly, every moment of every race, aware of the dangers involved. I was always worried, living in the fear that one day he would go out and never come back. It was as if every time he was going out to race, he was going to war – it was really that unpredictable. So, it was not uncommon for me to be apprehensive under the best of circumstances; every morning that he went out to ride, I would worry, and then when he would finish for the day, I would feel a sense of relief that he was OK, that he hadn't had a fall. I suppose because of Senna's death, and because Declan – usually so calm and unperturbed – seemed so shaken by it, I was even more anxious than normal that morning.

At the time of Declan's race, on the afternoon of 2 May, I was getting ready to go to Lester Piggott's yard – where Declan had horses – to ride out. The BBC was on – I would always watch

Declan's races on TV, of course in part because I was supporting him, but also because at the end of every race, I would feel happy that it was over. So, I was on a chair in front of the TV, leaning over it, ready to go and ride a horse that we kept at this yard – a horse that was being rehabilitated, that needed riding twice a day. I had on my breeches, boots, all of it. I was half watching, half ready to go, keys in hand.

And then I saw him fall.

I knew instantly that it was bad. I had seen Declan fall in the past on television, and usually it was a matter of minutes before he was back up on his feet and the commentator would typically say, 'Declan Murphy is about to get up' or 'Declan is winded' or whatever.

This time, there was nothing. I knew there had to be something very wrong when the BBC presenters on live TV seemed to hesitate and stall. Sir Peter O'Sullevan's usually very smooth, very confident voice had noticeable breaks in it as he cut off with a cryptic 'Unfortunately Declan Murphy is not yet up on his feet'. This was easily six or seven minutes after the race. Then I heard Julian Wilson come on to say, 'OK, we're going to leave it there. We don't have any more news of Declan Murphy at this time.'

And then, quite abruptly, they went off air.

I was convinced he was dead. I had imagined this moment many times since Declan and I got together, but nothing really prepares you for it when it happens. I sat on the sofa, frozen with the shock of it all. I didn't cry, I just sat there.

There was this silence that was deafening; that felt like it went on for ages before anything happened.

Suddenly, I snapped out of it. I couldn't just sit there. I felt like I had to do something, find out what was going on. So I rang Declan's phone, hoping that someone in the weighing room would answer it and tell me what had happened, where he was, if he was alive. I must have tried like an obsessed teenager – twenty or thirty times, at least. But his phone just rang and rang and rang. Eventually, I gave up. Next, I rang the racecourse, who wouldn't give me any information at all. Then it occurred to me to look up the nearest hospital to Haydock Park. I assumed if something had happened, a hospital would logically be the first port of call. I found out it was Warrington and I called them. Finally, I got something back. They told me that someone had indeed called from the racecourse and that they were expecting a casualty to be brought over by ambulance. They weren't allowed to tell me anything further, but this was enough for me – if they were bringing him in, it meant that he wasn't dead. Yet.

I had barely put the phone down when I got a call from Declan's driver, Jim Hogan. His usual big, bold, full-of-life voice cracked as he uttered the words that will be etched in my memory for as long as I live. 'It's not good, Jo, it's not good. Blood. There was blood everywhere. So much blood.'

After that, for the next few moments, it was the calm before the storm. I remember walking around the house, looking at pictures of Declan and me, our house, my life, thinking, It's all over.

Seconds later, I looked up at the TV and what I saw chilled my spine. Sky News had put out a 'Breaking News' headline that flashed across the television screen before my very eyes: 'Irish jump jockey fights for his life', it read.

And then the phone started to ring off the hook.

I kept screaming at the phone because it kept ringing and ringing, and all I wanted to do was find a way to get to Declan. Then, like a missive from heaven, a path appeared. One of the calls was from a friend of Declan, who told me that he had heard the news and wanted to help. In a gesture of unprecedented generosity, he told me that he could have his plane ready for me within the hour. I was over four hours away from the hospital by road on a good day; that day was Bank Holiday Monday and I knew the roads would be jammed. I could think of no quick way of getting up there and suddenly I had a plane I could use – it was like a miracle, really. What more could I ask for?

Jockey Michael Hills's wife, Chrissy, had always been a good friend to me and she offered to accompany me on the plane. There is a landing strip at the back of the July Course in Newmarket that runs parallel to the racetrack and this is where we headed. When we arrived on the runway, we saw Barney Curley standing there, and the three of us made the hop across from Newmarket to Liverpool in fifty minutes.

It was the longest fifty minutes of my life. Chrissy tried to comfort me as best she could, but there wasn't so much as a whisper from Barney. Not a shred of compassion.

The silence stung.

My boyfriend, my love, was lying in a hospital somewhere, in what state no one knew. I was a twenty-two-year-old girl confronting possible death for the first time in my life. Declan worshipped Barney. Under the circumstances, I would have expected a little bit of empathy, some reassurance, a scintilla of solace. Any human connection – a touch on the arm or a 'Don't worry, Jo, I'm sure he'll be fine' would have gone a long way in helping me ride the waves of anxiety.

But nothing ever came. Barney Curley sat in silence.

At the time I remember thinking how heartless he was, how inconsiderate, how absolutely selfish. Many years later, I would realize how wrong I was in my assessment, how inconsiderate, how absolutely selfish. Barney Curley's badge of distinction was that he never showed his cards. His face was always an impenetrable façade, devoid of any display of emotion. It was no different that time on the plane, as he sat by himself, his hard eyes closed, head leaning against the window in silence. But I should have known better, because even while his face betrayed nothing, his silence spoke a thousand words.

I realize now that this man, usually overflowing with self-confidence, was riddled with guilt. He had always had faith in Declan's class as a jockey, but it was far greater than that. Declan had been like a second son to him – Barney didn't love many, but he genuinely loved Declan. It seems inevitable now that he was contemplating his part in this whole thing – if he hadn't brought an eighteen-year-old Declan over from Ireland in the first place, we wouldn't be here, on this tiny six-seater plane, going to see if he was alive. And so if Barney was silent, it wasn't because he was indifferent to my emotion; it was because he was consumed by his own.

We all become selfish when we're floundering in the puddles of grief.

When I got to The Walton Centre, there were paparazzi all around. The entrance to the hospital was mobbed – there were cameras everywhere trying to take my picture, reporters trying to ask me things, even as I fought my way through them and made my way to the main reception.

As soon as I got inside, I found the first member of staff I could and I asked the two questions foremost on my mind: One, was he alive? And two, could I see him?

The answers, in order – because the order was very important – were, yes and no.

I was told that Declan had been brought in alive and that he was in the operating theatre at the moment; he had been in there for a few hours already. So I sat in the waiting area

– a windowless room, bright with fluorescent ceiling lights. The space was long and narrow, with three or four distinct seating areas, consisting of chairs arranged in a U formation around low wooden tables with fake flowers on them. There were signs everywhere – on walls, on stands, even stuck on the floor – in bold, bright letters, reminding people to wash their hands. I sat alone, staring at the bright purple plastic flowers in front of me, staring at the signs, thinking, Is he even going to come out? Is he going to come out in a sheet?

After what seemed like hours, I was received by a man who introduced himself to me as Professor John Miles, Declan's brain surgeon. He escorted me to a room, sat me down and offered me a glass of water. He then proceeded to inform me, in the most matter-of-fact tone you can imagine, that my entire world as I knew it had been ripped apart. Apparently, Declan had suffered a very severe head injury and was lucky to be alive. That, and the fact that two blood clots had been removed, were the only two bits of good news. The bad news came in more generously. First, there were no guarantees that he would live. If he did live, he was most likely going to be brain-damaged. He might be paralysed. One of his blood clots had been a hair's width away from his optic nerve, so it was more than likely to have been damaged. As a result, he might wake up blind in one eye. 'We don't have definitive answers at this point,' he had told me, not unkindly, 'but I am just preparing you for what might come.'

It's strange. You see dramatic moments like this in movies and you think it'll be very emotional, but actually it's all very practical. You don't cry; you think. The thoughts in your head are so functional, so objective, they border on seeming insensitive, cruel almost. The right side of the brain controls the left side of the body, I remember thinking to myself. John Miles had said he might be blind in one eye. Which eye, I wondered, left or right? Had it been damaged by the surgery or by the accident itself? If it was the surgery, it was likely that the right eye would be affected. But if it was the accident, he'd probably lose function in multiple organs on the left side of his body, including his left eye. Yes. Left eye, I concluded, he's probably going to be blind in his left eye.

Then, suddenly, I suppressed a desire to scream. What the hell was I doing? 'Stop it, Jo, stop it,' I said out loud to myself. 'Get a grip, for God's sake.' So I stopped myself from going any further because I knew that even one more step down this track meant that I stood in real danger of being derailed. I had to stay in control, because nothing else seemed to be in any sort of control. So instead, I just sat there in silence.

And then I was taken to the ICU to see him.

It really was very theatrical. They walked me there wordlessly and then, in one sweeping motion, they drew back the curtain and left me to process the spectacle that lay on the other side. It took several sharp intakes of breath to compose myself, because the man on the bed looked like a monster. His head was huge and distended. He had large rings around his eyes that were jet black. Funnily enough, there wasn't a mark on his body. It was wired up, but it looked absolutely normal. It was just his head and his face. And his eyes. What wasn't covered in bandage was as black as night. I kept staring at that face, thinking, Who is this? Did they really want me to believe this was Declan? It wasn't him there. I didn't know this person.

I still don't think I cried. Shock, I learnt, kills tears.

I kept thinking about the eyes. Those eyes on that body. They weren't Declan's eyes. I knew Declan's eyes. His eyes were what first drew me to him . . .

I first saw Declan on Newmarket Heath. I used to ride past him when I was in a string of horses. I was in one string and he was in another. You would get these massive strings of horses that would cross each other at certain times in the morning – sometimes there would be five, sometimes ten, sometimes twenty. We used to have to ride through the village of Exning to take us up to Newmarket Heath, and there were areas where the paths narrowed and the strings of horses would pass each other, single file, at very close range.

The order that you rode within the string was determined by hierarchy. Jockeys, being the best riders, would always lead, followed by the rest of us, riding for fun or as a hobby or for a love of horses. Declan would always be at the head of his string, and I would be somewhere in the middle of mine. And every now and again, we would make eye contact. There was always something different about Declan – he looked different to everyone else in the manner in which he rode a horse; he was always very relaxed, very sure of himself. But the most distinctive thing about him was his eyes. Perhaps because that was all I could see. He wore green silks on his helmet with this white shamrock, and a scarf that covered most of his face, but I could always see his eyes. He had an intensity about his eyes.

And now they wanted me to believe that the man lying in that room was Declan. I wanted to scream – those aren't his eyes. But I knew nobody would listen.

I held his hand then – it was soft and warm. 'Hello, Dec,' I said. But of course he didn't say anything back. For the first time that day I started to cry. 'It's me, Dec,' I whispered. I was trying not to say too much because I didn't want him to hear me crying. I was allowed ten minutes with him before the nurses took over again.

From that point in time, I lost track of day and night and date and time. I sat by his bedside and I never moved from there. I became a nurse's nightmare, obsessed, watching those machines, the colours, the lines, the mechanical sounds. If anything changed – even the tiniest variation in the colour or shape on those little screens – I would panic. I think the thing that strikes you the most in the ICU, when anyone is on intensive care, is the ventilator – it's that noise – when you know that they're *not breathing but that* it *is. I remember this one time when it went whoosh, whoosh, then did a longer than usual whooooosh and then beeped. I started screaming for help, 'Nurse, NURSE? Is anyone there? I think he might be . . . dead?'*

I had to make myself indispensable. It was the only way to cope.

When I was too exhausted to sit, I would lie on the floor. I would close my eyes, but not for long, because I thought, If I sleep, if I go, the whole thing goes. I did this for days.

I spent my time talking to him as if he was normal, about very key things in our life at that point. I had to force myself to sound positive, really focus on not crying, not sounding worried. I kept thinking, What can I say that will get him to wake? What can I say that will trigger something in him? Because I wanted him to come home now. I was done with this. I wanted it over. I wanted the love of my life back. So I was doing everything I could – reading, holding his hand, talking. I was convinced I could wake him up, that's how naive I was.

It was four days after his accident that he came out of his coma. Everyone had pretty much given up by that point and they had contemplated turning off the life support. They had tried three times to bring him out of it but it hadn't worked. I witnessed it all. I was in the room and the way in which they tried to take him out of it gives me nightmares to this day. No one prepares you for this kind of thing. There's no sugar-coating the way it's done. They just literally took the oxygen away and left him to breathe on his own. And he couldn't. And I was there, the whole time, watching it, watching a man gasp for air. I could see the fear in his eyes. It was savage.

And then, when everyone had given up hope, he woke up.

He woke with the broadest Irish accent you've heard in your life – as if he was a child who had never been taken out of Ireland. And he remembered all the facts of being a child. It was really quite surreal. They asked him all these questions and he got them all wrong. Not wrong, exactly – just misplaced in time. It was like they say: when you die, your life flashes in front of you. In his case, it was as if he was dying and his life was flashing before him, and then suddenly he woke up when he wasn't meant to. The reel stopped moving. And he was still a child.

The whole thing was completely bizarre. I was thinking, Wow, apart from the fact that you're now alive, you're speaking with this accent that I don't recognize, about these things, these people that I don't relate to. Who are you?

I'm not sure he knew either because he did some really strange things after that.

One time, I think it was Day 6 after the surgery, when he had been moved from the ICU to his private room, he woke from sleep, looked at me, and said, 'OK, we're going to Browns now,' in a very definitive but nonchalant way. Then he tried to sit up and found he couldn't support himself, so he fell back down. I was in the room with two or three nurses at the time and we all looked at each other. Here was a man, completely tubed up, his head bandaged, looking horrific, and he wanted to go to Browns?

It is hard to imagine the chaos within him; how confused, how disorientated he must have been inside his own head. I didn't want to add to it all. We used to go to Browns together every Saturday night for dinner, it was one of those things we did ritualistically. I was amazed he even remembered Browns because he didn't seem to remember much else. Anyway, I thought I should just play along. So I said, 'We'll do that later, darling, we'll go later.'

Then he pulled himself up and said, 'No. We're going now.'

He was forcing himself to sit upright but he wasn't physically able to, so he was swaying in bed – wires and all – like an unruly drunk.

I sat down beside him, tried to calm him. I said, 'Yes, sure, we'll go in a few hours.'

He got quite animated at this point, literally yelling, 'I've got to go, I've got to go. I've GOT TO GO NOW.'

And then he tried to swing his legs over the side of the bed as if he were trying to stand up. But he couldn't feel his legs, he had no sensation, so he couldn't do it. The more he couldn't do it, the angrier he seemed to get. When the nurses approached to help him back into bed, he stuck his arm out, trying to stop them, saying, 'NO, it's OK. Jo's sorted it.'

The nurses said, 'Declan, sit down now, please.'

I said, 'Dec, we can't go tonight, they're fully booked.'

He was shouting now, at the top of his voice, 'NO, NO, NO, we are going NOW.'

He was yanking at his tubes like a madman, so out of control that the nurses had to physically restrain him.

Then he swore. He never swore.

'Who are these fucking people? No one tells me what to do. Jo, can you just sort it, please? Just sort it.'

I couldn't sort it. I wasn't God.

I left the room and locked myself in a bathroom and I cried.

Then I laughed. I laughed at myself; at my naiveté. What did I think would happen? That everything would just go back to being the way it was? When you open someone's head and you put it back again, how can it ever be the same?

It was about a week later that Declan got it into his head that he had recovered. He didn't want to stay in the hospital, he said. He didn't want care and he didn't want the drugs. When he discussed this with me, he wasn't asking for my opinion – he was telling me. I had never heard of anything more ludicrous than a man who had just had brain surgery suggesting he didn't need medical assistance, but I didn't argue. I didn't try to change his mind. I went along with it.

I did this for three reasons. One, I was just beginning to understand the level of distress he was facing and I wasn't about to add to his turmoil by dissenting. Two, I knew that when Declan was determined to do something, he would do it – he had been that successful in his career, only because of his grit, his steadfast determination. It was his way of being, and there was no changing his mind. Three, at that point in time, I was the only person he was allowing to even come close to him. I felt it was my role to stand by him, to support him, to trust him. And I did trust him. Declan had always known best what was right for him, even in the most difficult situations – I had to believe that it would be the same this time.

It wasn't as easy to persuade the doctors, who were unequivocal that without medical care he stood no chance of recovery. But he argued so convincingly with his surgeons – this was a man who had won three racing appeals against the implacable stewards of the Jockey Club – that they finally relented. The team at The Walton Centre was so astonished by Declan's boldness and his inexorable self-confidence that they shook their heads in amusement and asked me repeatedly, 'Was he always like this?'

And so it was less than two weeks after the accident that we were brought home in an ambulance. It was a very poignant moment for me because when we got home later that day, and the ambulance left and I closed the door behind us, all help as I knew it stopped. It was just me and him. I'm still not sure to this day how we managed. It was all very organic. It wasn't dissimilar to coming home from hospital having just had a baby, someone handing you the baby, saying, 'There we go, that's yours, look after it now.' And you, not knowing anything about what to do, where to even begin, thinking, What am I going to do with this baby?

You make it up as you go along.

So we took it step by step, day by day. Time, light, darkness, day, night – it was all the same to him. He slept a lot, mostly during the day; often, he would fall asleep in the middle of a sentence, or just sitting on a chair. At night, he lay tossing and turning until the medication took over. Many times he would wake from that unnatural, induced sleep, drenched in sweat, screaming uncontrollably. When I asked what was wrong, he wouldn't want to talk about it. 'Nightmares,' he would say, and then change the subject. I didn't push it. I didn't show any anxiety. I just pretended – like he did – that nothing had happened.

There were days when it was very hard. He had always been a strong, proud man. No matter how dire the circumstances, I had never seen him ask for help or support. Seeking help meant accepting your own vulnerability. And he viewed showing – or even recognizing – vulnerability as a weakness. He had once said to me, 'No sane man would ever want to have a weakness exposed.' And yet, those days, he couldn't do a thing alone. He needed help to stand, to sit, to lie down, to get up. One of the first things he wanted to do when we came home was to shave. I tried to help him but he wanted to do it himself, even though it took him three hours.

He was very weak, because he couldn't really eat. He could manage soft, mashed-up foods; he survived on fruit and water, really, for days on end. I had to help him bathe because I didn't want him to slip on the wet floor. I know he hated it, and there were times when he would look away, as if he couldn't bear to be there, as if he couldn't bear that I was there, couldn't bear the fact that our life had been reduced to this. Many times, I'd feel an indescribable sadness – for him, for me, for us. But we had no other option – he was still very, very injured. He had his bandages. He had the stitches. There was dried blood everywhere. Half his head was shaved. He had the long, curving 10-inch scar of the incision. Then there was the dent. It was horrendous.

One day, he asked me to put on the video of the accident. I said, 'You can watch it if you want, but I'm not going to sit here with you.' I walked out of the room; he watched it alone. People said that I must be a very strong person. I didn't see it like that – from the moment I saw Declan fall, someone else was in my body. But as painful as it was for me, I cannot begin to imagine how difficult it must have been for him to come home. On some level, I'm sure it gave him comfort – at least in spirit – but on another level, it must have been hugely traumatic to cope with, because the balance of things had changed so drastically. Everything around him was the same – our house, the furniture, his clothes, our dogs, his pictures on our walls – but he wasn't.

Ironically, what added to the whole thing was the fact that suddenly we had all this media attention. I was a professional model by now; he was a famous jockey, but somehow, Declan's story seemed to capture the imagination of the world outside racing. My phone was ringing constantly – can we do this TV interview, can we broadcast live, can we have cameras when you arrive home? It was mad, really. A complete whirlwind. He had been elevated to celebrity status overnight, or we had, together, I suppose as a couple; because of our story, because it was one of triumph.

But I don't know if it was that straightforward. He had survived, yes, and for that I would always be grateful. But he had changed. I never quite knew what he was thinking, what was going on inside his head, what he remembered, what he didn't. He never spoke about it. He never spoke about the accident. He never spoke about how he felt. He became quiet, preoccupied, withdrew into himself. How do you analyse the psychology of a man who will reveal nothing of himself even to those closest to him? You don't. You accept. You adapt. I did both. I was a new Joanna, he was a new Declan. Someone had drawn a line in the sand and stupidly, unknowingly, we had crossed it.

When I was in my coma, I only heard one voice – it was Joanna's. When I awoke from my coma, I only saw one face – it was Joanna's. No one expected me to survive. But miraculously, I did. And so, in theory, for Joanna and me, you may believe that the hardest part was now behind us, that there seemed no reason why we could not rise from the ashes and continue to live the life we had lived before, together for ever.

Except life is rarely that simple.

Sometimes love stories have happy endings and sometimes they don't. I wish I could tell you otherwise, but here's the grown-up truth about love – it's not always a fairy tale. I have thought about this long and hard, perhaps more so now, than then; the benefits of hindsight and perspective are a splendid thing. And here is where I come out. Love sustains life, there is no doubt about it, and that imbues it with an immortality greater than life itself.

In the depths of the crevasses of a glacier is an ice of a crystal blue, with a purity and a clarity that is startling. And yet the outer surface of most glaciers appears black from the residue of volcanic ash, or the stones and rocks, that have settled on it over time.

We are an imperfect people in an imperfect world.

And sometimes, even love is flawed.

When I woke from my coma, I didn't remember Joanna in the way I was meant to.

I knew her, I drew more comfort from her than from anybody else, but she had morphed into someone from my childhood.

I loved her, but she felt like a sister not a lover. I was twelve years old in my head – what did I understand of romantic love?

Of course, while I had gone back into my past, she had stayed here in the present, in the same place that I had left her. And when I came back, my present had become my past. The unfairness of this consumed me. I ached to bridge the chasm, but I didn't know how.

And while I was acutely aware of the turmoil in my head, I couldn't make sense of it. I couldn't bring myself to talk to her about it. How does one swallow the shame; how does one summon the courage – to do this?

But you can't pretend feelings. You cannot live a lie. She deserved more than that, and what she deserved I couldn't give her – that privilege was taken from me. When the surgeon cut open my skull to save my life, he also did something else. He cut open our lives and tore out pages of the story that should have been ours to have.

There is an enormous sense of loss when you're meant to feel something that you just don't feel. I cannot be so bold as to presume what Joanna was feeling when we went our separate ways, but the pain I felt was unlike any other kind of pain I had felt before. It threatened to drown me in a sea of my own incomprehension, leaving me with an emptiness so vast, I couldn't even see the shore.

Nobody understood the magnitude of what Joanna did for me in those black months following my accident. I don't think she fully understood it herself. Sometimes you don't realize the impact of what you are doing, because you are just too busy doing it. When I came back from the hospital I was as helpless as a baby. I was starting to pity myself. I was starting not to care. I was starting to give up. Had it not been for Joanna, I could not have risen above that.

Joanna fuelled my inner strength to such an extent that I convinced myself that no adversity was going to bring me down. Adversity was not going to affect me. I genuinely believed this. She ensured that even when I was in my lowest place, my confidence was intact. She stood by me when no one else did, at the darkest moments of my life, and, if not for her, I would perhaps have never seen the light. And try as I may, I cannot fathom how it must feel to be consumed like this by the responsibility of someone else. Joanna was by my side when I took my first gasping breaths without a ventilator; she held my arm when I had to learn to walk again; she consoled me when I fell, beating my sticks against the ground in frustration. My sorrows were her sorrows; my joys, hers. Every step I made towards recovery was not just my achievement, but hers, a collective victory that ought to be celebrated. Another dawn breaking was one more day further from my fall.

I never had the chance to express the full extent of my gratitude to her. I don't have a good reason for this. It's very easy to get lost in yourself. But I say it now – my thank you – and I say it in a tide of emotion so strong it almost sweeps me away.

Sometimes, I cry.

I cry from within.

In gratitude and grief.

In happiness and praise.

In honour and tribute.

I cry from within, for the wonder of you.

Thank you, Jo. My life, I owe, to you.

There is no book without love. Because there is no story without love. My story had love and her name was Joanna.

Mr Turpin and the Grey Stallion

If there is only one thing you take away from my story, let it be this:

The first step towards conquering adversity begins with the will to do so. It starts from within. You have to want to. You have to need to. You *have* to. That's all. There is no disqualifying the sins of our past. But to dwell on them is to delay the ability to rise above them. Fairly or unfairly, we might find ourselves on the receiving end of forces outside our control, but we have total control over how we choose to move beyond them. We have to take charge of our own lives. We have to take responsibility for our actions. We have to make decisions that we believe in and then, no matter how hard it might be, we need to own those decisions. There are no excuses. The dog never ate it.

I don't think anybody knows the power within them until they are thrust into a situation when that is all they have to depend on. I had nothing other than the power within me that said if I willed myself strongly enough, if I shouted loudly enough, if I believed deeply enough – that I *could* do it, I *would* do it.

*

Pat Hogan was the king of the point-to-point world in Ireland, an internationally renowned bloodstock agent and a very keen huntsman.

My family had been associated with Pat Hogan for a long time. My uncle rode with him, my brother Pat started riding with him and then later, when Eamon decided he wanted to be a jockey, he would go up to ride Pat Hogan's horses on his weekends. Often, I would accompany Eamon on these occasions and that's how I first met him. It was Pat who got Eamon a job with Kevin Prendergast on the Curragh, and it was because Eamon was riding for Kevin Prendergast that I went up to the Curragh in my summer holidays. This is where I would meet Kevin Prendergast myself and then a year later, at his insistence, get my amateur licence and go on to win my first big race on Prom. So perhaps, if you think about it like this, my story as a jockey really started with Pat Hogan.

Once, quite early on – I must have been about seventeen at the time – I happened to be at a point-to-point race at Kilmallock, where I witnessed quite the peculiar little sight. Pat Hogan's horse, Mr Turpin, was running the race and instead of turning the bend, he went straight on – he actually galloped off into the countryside with a terrified jockey on his back!

It was an experienced jockey riding him, but Mr Turpin was feral. He shot like a bullet through the wing of the fence, and ran around the field like a horse in the wild. No one was able to calm him enough to bring him back.

They say history never repeats itself, but the following week at Thurles, Mr Turpin gave the historians (and the bookies) a proper run for their money – quite literally speaking. This time, the unfortunate rider aboard Mr Turpin was Pat Hogan's jockey, Enda Bolger. To the dismay of the jockey, to the frustration of Pat Hogan, and to the amusement of the watching crowd, Mr Turpin pulled the same trick again!

I, like everybody else, wondered at the bold waywardness of this horse.

One week later, for a reason I never understood, Pat Hogan thought I might want to ride Mr Turpin in a point-to-point at Nenagh. Why he thought I would have any luck with the prodigal horse when far more experienced riders had failed, I wouldn't know until later, but I was thrilled.

I had only ever ridden in three point-to-points before then, but how could I turn down such an adventure? Pat Hogan was an institution in himself and the excitement of riding for the king of the point-to-point world was an opportunity I simply couldn't walk away from. That it was the unpredictable Mr Turpin on offer made it all the more compelling.

I was in my last year at school doing my Leaving Certificate and under no circumstance would my mother have let me interrupt my studies to ride the race. So I took the easy way out and decided not to tell her about it at all! I was the youngest boy in the family, the apple of her eye, but this didn't preclude her from being extremely strict about my studies. School always came first. And so, as is often the case with these things, it's easier to ask for forgiveness than to ask for permission – and that's exactly what I did. Eventually, I knew I would find a way to win my mother back. I did, however, ask permission from my father – my willing partner in crime – and he drove me to Nenagh to ride the horse, perhaps even more excited than I was.

Notwithstanding the fact that 'fun' was my primary motivation for being there, I did still want to win, and I won't lie, it was tricky from the start. There were only six runners in the race, making it a challenge to keep Mr Turpin covered. Enda Bolger, the jockey who had attempted to ride Mr Turpin at Thurles, was in front of me, riding the favourite in the race, Moncai for Christy Kinane. At the first chance I got, I buried Mr Turpin behind Moncai, so he couldn't see any daylight. In my inexperience, I thought I had the horse covered, but as we approached the third last, Enda Bolger moved across the fence in the hope that my horse would unexpectedly get daylight and – predictably – run out through the wing.

But it didn't happen.

There is a saying we grew up with – A horse spooks at two things: one, things that move; two, things that don't.

Mr Turpin was clearly an unsettled horse, easily spooked, and a scared horse is a dangerous horse. But if there was anything I had learnt from my pony-racing days, it was that the first step towards increasing a disillusioned horse's confidence is to manage your own emotions – if the rider is afraid, the horse will be doubly afraid.

At the time of the race, I was an inexperienced schoolkid – I knew nothing. And on this occasion, my ignorance gave me a strange advantage. I didn't feel fear – mainly because I didn't know enough about fear to feel it. And as a result, Mr Turpin felt no fear. We fed off each other – I was calm, so he was calm; he was calm, so I was calm.

And it was with total composure that I held the horse together and, turning into the straight, snuck up Enda

Bolger's inside. I steered Mr Turpin decisively up to the last fence, jumped it, and won.

Later I learnt that Enda Bolger had flat out refused to ride Mr Turpin at Nenagh. *Our experiences shape us; they either give us faith or they take it away.* Enda Bolger had been clouded by his past experience, something Pat Hogan knew I hadn't been touched by. He was betting on the fact that I was fresh and enthusiastic, and that the self-confidence I had would transmit to the horse.

It wasn't until many years later that I heard from Ned Mitchell, whose father Cecil Mitchell's horse Rich Hill was also running in the race, that he, his father and Pat Hogan had been standing at the first fence, discussing their chances. Given Mr Turpin's past bad behaviour, Cecil Mitchell had been certain that his own horse would win. But such was the conviction Pat Hogan had in me – in my self-confidence – that I had barely jumped the second fence when he said definitively to the others, 'I have a right good young jockey riding my horse today – he won't run out.'

For someone who never wanted to be a jockey, for someone who never thought about it very much, I learnt three invaluable lessons from that experience. One: when you are out there fighting, competing, on a battleground, in a race or *for your life*, your destiny is your own; you only have yourself to depend on. Two: my rival in that race had tactically tried to expose my inexperience on that horse, but what I lacked in experience, I made up for in belief. And belief would be the rock in my darkest hour. If I believed I could do it, I did it. Three: even the most highly strung of horses can rise to the occasion, but only if they don't get disillusioned. It is *exactly* the same with humans.

*

My first act after emerging from my coma was to get myself out of hospital. I was in unchartered territory with the physical and mental consequences of the surgery, and as supportive as everyone in the hospital had been to me, the environment was adding to the layers of confusion I felt. I didn't like being looked after. I didn't like people rushing to my aid when I faltered. I didn't like the unintended – but unavoidable – looks of pity I got around the clock. I didn't like being a patient. It spooked me. And just like Mr Turpin, if I was spooked, if I was afraid, I stood in danger of running through the wing of the fence – and not being able to be brought back.

So I had to leave. In my mind, I was in a very happy place in my childhood. In the hospital, I felt naked, vulnerable, exposed. Totally stripped of control. It didn't fit. It didn't feel right. Nothing made sense. I wanted to recover, but only if I was not treated like a patient would I stop being a patient.

When I said I wanted to leave the hospital, it came as a huge shock to people. Except for Joanna, everyone else dissuaded me, arguing that I was being premature, that I would not be able to manage on my own outside the fully equipped hospital environment. I was very sick, the doctors said, and my recovery would be slow and painful at best. Especially if I left the hospital. In the hospital, where I was being monitored constantly, maybe I would have a chance.

In the end, I convinced them to let me go. But I could not have done it without Joanna, because she was the support vehicle that I so desperately needed. They were right, it wasn't something I could do on my own. You can take the decision on your own that you aren't going to accept the situation you are in. You can take the decision on your own not to stay in hospital. But you need someone to support you. And Joanna was the person who supported me. Because she knew what was right for me.

And what was right for me was to be in control of myself. Yes, I had just suffered major trauma. I had undergone brain surgery. I was incapacitated, I was paralysed, I was damaged. But it didn't change the person I was. At my core, in my DNA, I was a sportsman, and when sportsmen have a goal, they need to achieve it, obsessively and compulsively; it dominates their existence until they do.

I had a goal, and this time it was personal.

My goal was my recovery, and no matter how impossible it seemed, I had to get there. It would be my hardest challenge yet, the ultimate test of injured mind over matter, but I would get there. I had to. I was a jockey – the further away the winning post, the greater the drive to cross it.

I found myself consumed by this, by the desire to go back to being the person I was. And I believed I could only achieve my goal if I was able to put myself in the place I needed to be, in my own head. So the way I did this – the only way I knew how – was to deny there was anything wrong with me. Just like when I was on Mr Turpin, I denied fear, I denied any weakness that threatened to dilute my focus to win, so now, I denied my condition.

But first I needed to be in an environment that allowed me to regain control of my own mind. Which meant leaving the hospital.

In the end, I was only allowed to leave on the proviso that I was accompanied by a fully trained nurse and would be under her constant supervision. This arrangement did not last long either, and I decided I didn't want the nurse. None of this is meant to sound ungrateful, and try as I may, I cannot explain the reason behind it all, except for the fact that my mind wouldn't allow somebody else to take care of me. I didn't want to feel like a sick person. I didn't want anyone doing things for me. I wanted to be responsible for myself. I wanted to do everything myself. And I wanted Joanna. Joanna was not somebody else; Joanna was an extension of myself.

I wanted to feel normal. With Joanna, I felt normal.

I wanted to feel safe. With Joanna, I felt safe.

So we left the hospital and came home.

But if I was under any illusion that coming home meant that I would miraculously regain so much as a whisper of normality, it was quickly shattered. No sooner than I thought I had it, I lost it.

When I was a little boy, and I wanted something badly, I would close my eyes tight, scrunch up my face and will it to come true. And I truly believed it would. We marvel at the innocence of children, but sometimes, young or old, child or adult, this naiveté is what keeps us going. When you want something *so* badly, you start to believe it. You forget that dreams are nebulous.

And then you wake up.

I had unmitigatedly underestimated how difficult it was all going to be. Of course, it was what I wanted, and if I had remained at the hospital, I would have never recovered – this much I am certain of. But coming home was a sharp, unrelenting pain. It was home, but I didn't remember it. I was forcing myself, trying to visualize everything around me as familiar, yet nothing seemed the same. All my things were there, all exactly as they would have been when I left home on the morning of 2 May, but I didn't remember them. They could have been my things, they could have been anybody's things, I wouldn't have known the difference.

I remember looking around my house, this gallery of memories, looking at the framed photos on the walls – various pictures of myself in various races, my winners over the years, my moments of glory. I wanted to feel something when I saw them. Pride at my achievements? Sadness that it had all been taken from me? But I felt nothing. Because I didn't remember any of it. There it was, lining the walls of my house – my life in a snapshot. And I didn't remember it.

When Joanna brought me home, I didn't sleep for twelve nights. I was still heavily medicated and I would fall in and out of consciousness, but it was the restless, disturbed slumber of a restless, disturbed man. I didn't know the difference between day and night, light and dark. Minutes became hours, then days, then weeks. I wasn't angry, I wasn't sad. I was something much worse – I was indifferent. And this is what scared me more than anything else. If I was slowly but steadily descending into a state where I was devoid of emotion, it meant that I was losing control.

One time, my friend and financial adviser Kelvin Summers came to visit me. Joanna and I were renting Bob Champion's gate lodge in Newmarket at the time. I was sitting in a chair, facing him, but even while he was speaking, I was staring out of the tiny window, looking outside, my mind clearly somewhere else. Kelvin stopped talking then and asked pointedly, 'Declan, what are you looking at?'

'The tree,' I said. 'It's my lifeline.'

He got up and peered out of the window to see what I seemed so engrossed in. But a few seconds later, he turned back to face me and, looking puzzled, asked me the same question again. 'I don't understand. It's an oak tree. What are you looking at?'

'I am looking at the movement of the leaves.'

'Why are you doing that?'

'It's what keeps me going, Kelvin.'

I was a top and I was spinning.

Spinning, spinning, spinning, spinning, spinning . . .

I sat for days studying that oak tree in the garden, just letting my brain run wild, which is what happens when you dream.

I WILL remember. I WILL walk. I WILL ride.

Will I remember? Will I walk? Will I ride?

I WILL. I WILL. I WILL.

Will I? Will I? Will I?

Listless days. Endless nights.

I didn't talk. I didn't shave. I didn't eat. I didn't smile. I forgot how to laugh. Sometimes my eyes would slide out of focus and so would my voice – even talking for more than a few minutes left me exhausted. I spent my days asleep or lying in bed or sitting lifelessly in that armchair, numbed by the painkillers, staring at the movement of the leaves.

Sometimes I tried to count the leaves. But after a while

I couldn't, I got confused. You don't think about it, until you start doing it, but it is a very difficult thing to do . . . One branch merges into another, you lose count, you start again . . .

I was drifting rudderless in open waters, a lost soul.

I had no purpose, no meaning. *I may as well have died.*

And then there was a turning point in my life. And it came, as these things do, in the most unforeseen of ways.

My friend and fellow jockey Ross Campbell came to visit me at the gate lodge. He had been at Haydock Park on the day of my accident in the jockeys' room when Joanna had tried to call my phone. He hadn't had the courage, he said, to visit me at the hospital, and so it was the first time he had seen me since that fateful day. I was sitting in my armchair, my body barely filling a third of it, my head half shaven, still heavily bandaged. When he spoke to me, I was able to respond, but my head didn't move; I had no expression. Then, when he left the room, I heard him speaking to Joanna outside. He was trying to keep his voice low, but I heard two words that triggered something deep inside of me. The two words were 'never recover'.

This is the thing about pain: you can try to disguise pain, you can try to disguise heartache, but your eyes never lie. If you've got pain, it will show in your eyes. I had pain and Ross had seen it in my eyes. Listening to his words, the utter resignation in them, did something to me. It was probably the first time since I had woken from my coma that I started to think that I was different, that something was wrong with me. Because you only see yourself. And I didn't see myself in that position of despair. But when somebody who knows you sees you and feels a sense of pity for you, loses hope in you, it wakes up something within you. Something that you thought was dead begins to spark life.

People had always remembered me for my intelligence. And with intelligence comes emotion. I knew instantly why I had become emotionless. And I knew what I had to do. The thick fog of confusion that had engulfed me lifted suddenly.

I had been here before – just on the other side . . .

Francis O'Callaghan, who came from the same village as I did, used to run the quarantine in Ireland for shipping horses to Australia and New Zealand. I would go down to Castletroy on the weekends to help exercise the horses. Essentially, there was a big lunging ring, and the horses were made to run around the circle – on a rope with a head collar – clockwise and then anticlockwise for about twenty minutes each way, to keep them fit and healthy.

The summer of 1980 saw a dramatic new arrival at the stables. A big grey stallion with a dangerous reputation was going to stand in Australia in a few weeks. He had been trained on the Curragh, and his obvious aggression had earned him the nickname of 'Man-eater', as well as the need for two handlers at all times, even just to go into the stable, or to be fed.

One Saturday morning, when I had finished lunging all the horses in the first yard, as Francis had asked me to, I decided to make myself useful and work my way down the remaining stables. Eventually, I got to the 'Man-eater' and, even from outside, I could hear the restless pounding of his hooves inside the stall. I opened the stable door with the cavesson in my hand, my sole intention being to put it on him, walk him out and lunge him, like I had done with every other horse. Call it my innocence, or my ignorance – whichever you like – but I never stopped to think about the fact that the stallion had always been handled by two people; in fact, my fourteen-year-old brain didn't so much as process any danger at all.

Predictably, no sooner had I opened the door than the horse made a go at me. Head tossed, mouth open, he reared up aggressively as he saw me approach. I backed away carefully, but right at that moment, I noticed a discarded piece of thick rubber piping, lying on the ground outside the stable door. On a whim, I picked it up and put it in the stallion's open mouth, swiftly, before he had time to react. Exactly as I had hoped, the moment he bit down on the pipe, I saw a calmness descend over him. I slipped the cavesson on and, leaving the rubber piping in his mouth, I walked him out and started lunging him.

Soon enough, Francis O'Callaghan and a few of the stable lads came around the corner. When they saw me, they ran towards me. Francis screamed, 'What the fuck are you doing? You're going to get yourself killed!'

I said quietly, 'I'm fine, he's fine, we are both just fine.'

And they stood there in disbelief, watching me, not comprehending what they were witnessing. Nobody, not even experienced trainers, had been able to handle this horse, and there I was, a fourteen-year-old kid, lunging him coolly without so much as a whimper.

When we were finished, I took the stallion back to the stable and, leaving the rubber piping in his mouth, I took the head collar and the cavesson off. Right before I left, I took the pipe out of his mouth, slipped it in my pocket and lo, there he was again, the 'Man-eater' in all his snorting rage.

I was at an influential stage of my life – that delicate transitional phase between childhood and adulthood – when perhaps one is able to see things with greater clarity than many adults. I could sense an insecurity in this horse. I didn't know why he had it, but something was making him afraid. And as is often the way with man or beast, the insecurity was manifesting as aggression. He simply wanted the world to know that he was in charge. And I was saying to him, 'There's no need, I'm not going to fight with you.'

I went back to the quarantine yard a few more times subsequent to this event and, with the rubber piping as my aid, I was able to handle the stallion each time without incident.

Eventually, when the quarantine period ended, the horse was due to travel to Shannon Airport to board his flight to Australia. Dave Bartle, chief flying groom of the British Bloodstock Agency, had flown in from England, and was to accompany Francis on the plane. Given my rapport with the horse, Francis asked if I would be willing to take a day off school to come along in the horsebox to the airport and then drop the horse off to his stall on board the plane. I agreed, and that morning I prepared the grey stallion for travel by fitting the rubber pipe in his mouth and tying it to his halter with two pieces of string.

When the horse was safely on board, and before I left the airport that morning, I convinced Dave Bartle that I believed it would be advisable for the rubber pipe to be left in the stallion's mouth for the duration of the flight. To his credit, he agreed readily, albeit with some amusement at this teenager looking up at him, giving him advice with such surety of mind.

However, I was to learn later that, during the flight, the string on one side of the halter securing the rubber pipe had come undone, leaving the pipe hanging out of the horse's mouth. One of the grooms had inadvertently removed it, and the horse had lashed out. In turn, the groom had reacted in instinctive anger. Things had come to a head – the stallion had retaliated in fury, swinging his leg out over the top of the stall and trying to kick it down maniacally. Forty thousand feet above the ground, flying over a world without borders, he had become completely delirious and nobody had been able to control him. Ultimately, for the safety of everyone else on board, he became collateral damage. They had no choice. They were forced to put him down.

I tell you this story for a reason.

Yes, I had been able to calm this horse with my rubber pipe, but in essence, what I had done was to lull him into a false sense of security. As long as the rubber pipe was in his mouth, all was well; the minute it came out, the horse reverted to his true state – one of deep-seated insecurity. Even my best efforts to allay his fears, while effective, were but ephemeral. Somewhere, somehow, long before I had met him, his spirit had been irretrievably broken. In the end we failed – we all failed – to save his life.

It is more complicated with man than it is with beast. There is a profound gap that separates us, and that lies in our ability, as human beings, to introspect. We are aware of our existence, beyond a simple recognition of self – we are able to affect our own destinies. We behold the world around us; we evaluate, we admire, we criticize or we simply remain bystanders. And in doing so, we understand our role in the world and we contemplate how we want to belong to it.

Animals have a survival instinct, but we have a deeper consciousness – we know that we are alive, we ponder the meaning of life, and we know that one day we will die. And so we have an energy, an enthusiasm, a spark to live life to its fullest. Unlike animals, we have the power of perspective, of self-reflection. We have the power to take that leap of faith to step outside of our bodies and see ourselves from a parallel perspective. We have the power to look within our core. We have the power to understand our deepest traumas. And we have the power to overcome them.

Medication, for me, was the equivalent of that rubber pipe. This is what medication does. It neutralizes your senses because it is trying to kill the pain, but in killing the pain, it also kills a lot more. As long as they kept me on the drugs, I would have existed in a zombie-like state of induced normality. But in the process, they would have extinguished the spark within me, swiftly and silently, like a gust of wind extinguishes a burning flame. And when the drugs stopped – when that rubber pipe fell out – and I emerged from this unnatural, sedated state, I'm convinced I would have never recovered. My spirit would have been broken. The fight within me would have been quenched. They would have had to put me down.

So, my second act after emerging from my coma was to

get off the medication. I had been kicked in the head by a racehorse at speed, I had almost died, my skull had been cut open, my mind had been badly traumatized, I had lost my memory, I had no feeling in my feet, I had no feeling in my hands, I had no feeling in my fingers, but none of that compared with being numbed of emotion. Of standing by helplessly while the drugs slowly and silently stamped out the life-force within me. So, just as if I had stayed on in hospital, if I had stayed on the drugs, I am certain that I would have never recovered.

Joanna understood this.

When I stopped the medication, Joanna never asked why. She never questioned it, she never told anyone, she simply supported me. In hindsight, her compliance could well have been perceived as a great irresponsibility on her part, because despite what I thought I was capable of, I could well have been proved wrong. That had to be a brave call for anybody to make. And so, to have the strength of character to believe in me when I was at my most vulnerable – that showed the kind of woman she was.

In her mind, it was less important what I actually could or could not do. What was more important was that other people never got the chance to underestimate what I could do. She played up the strength within me; she protected me from everybody.

So while the will to recover was ultimately my own, it was Joanna who gave me the initial push towards hope, towards possibility. It was Joanna who kick-started my mind, who fuelled my belief, who made me feel invincible. She was the lynchpin of my faith. Without her, I may have forever been lost.

When I look back, I think how easy it would have been for me to give up. Everything I knew had been taken from me. It had taken three seconds precisely from horse hitting hurdle to hoof hitting head. Three seconds to end it all. I didn't know if I would walk again. I didn't know if I would ride again. I didn't know if I would remember again. I didn't know if I would love again. I didn't know if I would live again. So, yes, it would have been easy to give up. But I couldn't accept that. Most people who went through what I did are not alive to tell their stories, so I had a different cross to bear. I had been spared my life. Now, I had to save it.

This I learnt from the most unexpected of teachers, in the most unexpected of ways. Thank you, Mr Turpin and the Grey Stallion. Thank you for your infinite wisdom. You didn't say a word, but you taught me all I needed to know. You showed me the way; out of the tunnel of darkness, and into the light.

Light

Coping with your own death, when you are not yet dead, is a strange thing.

When I look back, I realize that the hardest thing for me wasn't learning to walk again, it wasn't living with the dent on my head where the surgeon had cut out a part of my skull, it wasn't the lack of sensation in my fingers or my feet, it wasn't the blinding headaches, or the blank phases that would come upon me suddenly and take me into black pools of isolation.

It was English grammar. Semantics. When people spoke to me they would say:

'You were a great jockey.'

'You were such an articulate man.'

'You were the most stylish of riders.'

Wonderful, kind words, meant from the heart. But the grammar was all wrong. Everything was in the past tense.

I was. I was. I was.

But here I *am*.

There was a fire within my soul. I was determined to get my life back. To meet these expectations that other people seemed to have of me. To be referred to once again as a 'now', not a 'then'.

I was hanging off the edge of a cliff, with fingers I couldn't even feel. And I could either hoist myself up to safety or fall into the dark, cavernous mouth of the valley below. Above me in the sky glowed a rainbow, its vivid, multicoloured hues beckoning me. I knew I had to crawl towards it, towards the light.

I had to heal.

Bit by bit.

Part by part.

Body, mind, soul.

Perhaps it seems too complicated to think this way, to try to isolate the intricate, conjoined workings of a human being in the way that I did – body, mind, soul. But I had damaged each one and I had to repair each one.

Bit by bit.

Part by part.

I had to break it down.

I had to step outside my physical self and analyse things from outside in. I had to simplify. Otherwise, the red rocks of the earth would have opened up before me and I would have found myself falling . . . falling . . . falling . . .

But there, just there, was the rainbow, waiting for me.

The way I viewed it, I had two challenges to overcome, woven together by a gossamer thread, complex, fragile: one, the damage to my physical self and, two, the damage to my mental self. At the core of all this, at the heart of the web, lay my memory loss – the four years that had been snatched from me. Those four years and everything that came with them – the battles won and lost, the triumphs, the disappointments, the laughter, the tears, the memories, the magic – were gone for ever; I would never get them back. I simply had no choice but to accept this. But alongside it all came another reality, much harder to accept – the fact that I couldn't remember my most recent self. I found myself wrestling with the most basic of concepts: self-awareness, self-image, self-knowledge, self-identity. I didn't know who I was, who I was meant to be. Inside me was a stranger I needed to befriend but I didn't know where to begin. Allow me to replay the conversation I would so often have with myself.

Me, to me:

Who am I? Who am I supposed to be?

They say I used to race horses.

So maybe if I can race horses again, I'll become me again? But how can I race horses when I can't even walk?

Suppose I learnt to walk? Suppose I learnt to ride? Suppose I learnt to race?

Would that then make me, me?

So, I thought, rightly or wrongly, that if I could get myself physically back to the man I had been, it would help me regain my sense of self. Or at least, some part of it.

In any case, the physical recovery was easier to tackle and that's where I started. Here's what I was contending with:

Before Haydock Park

Height: 5'10"

Weight: 66 kg

Body fat: 8%

The stuff they said: 'He didn't need to work at riding, it just happened. The rest of us had to work at it, we were ordinary – he was just good. And anyway, he wouldn't have had the time to work at it, he spent all day in front of the mirror, combing his blond hair, trying to look good. And to be fair, he did look good. He was a strong athlete, full of muscle, strong shoulders, strong chest . . . a supremely stylish rider. Declan just seemed to be cut from a different cloth. He wanted to ride better than anybody else; he wanted to look better than anybody else – and he did.'

After Haydock Park
Height: 5'10"
Weight: 36 kg
Body fat: What body?
The stuff they said: 'He looked like a little old man, so frail, sitting in an armchair that he usually perfectly filled. Now he was sitting like a ragdoll, taking up about a third of the chair. He was always so strong and to see this sort of strong body reduced to looking like a shirt on a clothes hanger was a great shock. His face had become completely gaunt, his head was shaved, all the muscle was gone, his whole diaphragm was gone; it was just skin on bone, his clothes were hanging off him, like they were draped there. He was just a decrepit old man and a pair of sticks.'

It was hard.
It was simple.
It was impossible.
It's never impossible.

I had to turn back time.

I had to discard my 'after' and regain my 'before' as quickly as it had worked in reverse.

I had to take control of my own destiny. And I had to do it now. Carpe diem.

But there was a paradox: even in my impatience, I was in no mad rush. There is a saying in Spanish, '*Vísteme despacio que tengo prisa*', that translates roughly to 'Dress me slowly because I'm in a hurry'. I was completely patient despite my hurry. I didn't intend to jump headlong into my recovery. Just like I never jumped headlong into my races. *Ride the race to suit the horse and not the horse to suit the race.* I was the horse, the prize was my life; the race would determine how I got it back.

I made a game plan. Actually it was more like a battle plan. I was going to approach this slowly and patiently, step by step, as if preparing for battle. And I was going to win. I made lists in my head and I went over them again and again and again. What I could do, what I couldn't do, what I *wanted* to do. What I had lost for ever, what I could potentially get back.

I consider myself a man of faith. As children we were taught the Serenity Prayer, we would recite it every morning at school. We would say it by rote, mechanically, as children do, not stopping to think about the words or what they meant. Now, it served as my salvation. I asked Joanna to copy it out and tape it to the wall facing my bed, so it was the first thing I'd see in the morning and the last thing I'd see at night. She wrote it out in bold green ink on a sheet of white card, in her typically neat, slightly slanting cursive. It said:

God grant me the serenity
to accept the things I cannot change;
courage to change the things I can;
and wisdom to know the difference.

Did I have this wisdom? I don't know.

All I wanted to do was walk again. No, I lie. What I really wanted to do was ride again. I'm still lying, because what I truly wanted was to race again, and if you really want to know the truth – what I wanted, more than anything, was to win again.

There are jockeys who will tell you that racing horses is all they've ever wanted to do. I'm not one of them. I became a jockey by circumstance and it so happened that I made a success of it. But it was always just that. It was what I did, not who I was. And yet, at this moment in my life, I wanted nothing more than to be a jockey, if only to regain some semblance of control, to prove to myself that, despite everything, I was still me.

The first step of my battle plan was to walk again.

There was a porter at The Walton Centre called Tony. His job was to wheel me around in my wheelchair, just to get me out of my room, just for a change of scene. All day long, he'd wheel people like me around. I rack my brains to this day and I cannot think of a job more giving of oneself than that which Tony dedicated his time to doing. Day in and day out he'd walk the same routes, along the same dreary corridors, wheeling people who were sick or injured or damaged. Sometimes these people were afraid. Sometimes they were angry. Sometimes they were just in pain.

And really, if you look beyond the physical act of pushing wheelchairs around, Tony's real role was to lift people from their lowest lows and make them feel better about themselves. I cannot think of a greatness, more great. The memorable thing about him – the thing which endures – was that he never treated me like a patient. He laughed with me, he joked with me, he asked me questions about my old life – who comes up with names for horses, which one was my favourite horse, did my back hurt bent over like that all the time, did all jockeys have pretty girlfriends like I did? And in doing so, he created a sense of lightness in my being. Because he did something I yearned for, as one yearns for the first blush of colour after the stark whiteness of winter – he treated me like an ordinary man.

His role was small, but his heart was big. When we speak of angels on Earth, these are them.

⋆

Before I learnt to walk again, I had to learn to stand.

It's hard to believe sometimes the stuff life throws at you. Yesterday I had been a twenty-eight-year-old hugely successful jockey. And now, I was teaching myself how to stand.

Tony helped me. He would hold my arms above my elbows and support me off the seat. No, it wasn't support, really; what he'd do was haul me off the seat because I had no sensation in my legs. Even though I willed myself to, I couldn't do much work. So I'd dangle there, my body in Tony's hands, just trying to balance myself, just trying to get the movement going.

We did this every day, sometimes several times a day. Sometimes my legs would give way and I'd collapse back down within seconds. On these days, he would get emotional, tell me to go away and rest. I think he did this when the shadows of doubt crossed his mind, but he didn't want me to see them.

I was unrelenting. Sometimes I'd be able to take a step or two. On these days, he would laugh and I would laugh with him. Because you *have* to laugh. These were the great days. When I took ten steps off the wheelchair for the first time, Tony still holding me, his arms doing the work of my legs, he let out a whoop of joy. 'My God,' he laughed. 'The race-rider is walking.'

It was a start but it was not enough. I could stand. Now I wanted to walk. Fuck, I wanted to walk.

It was harder than I thought.

The house we were renting at the time in Newmarket was on Hamilton Hill, a private road with speed ramps along the length of it. I was taken for walks every day by Joanna, supporting me, holding my two arms while I attempted to balance myself on my walking sticks. We started off going for five-minute walks. By four minutes, I would feel shattered. Joanna begged me to be patient, to slow down, but I insisted on pushing myself, on doing more than I could. Sometimes, when family would come to visit, I would go for walks with them. My brother Eamon remembers having to physically lift my legs, one at a time, over each speed ramp – I couldn't even get my foot that high. This was three months after the accident.

One time, I was on my walk with Joanna, and as we were approaching one of the ramps, about 100 metres away from the house, I looked up and saw a car coming towards us. In reality, perhaps the car *was* travelling a little faster than it should have been, not slowing down far enough in advance of the speed ramp, but certainly nothing too out of the ordinary. However, so damaged was my perception of distance, so damaged was my brain, that I thought the car was coming straight at me, that it was going to run me over, and I panicked.

My brain was signalling frantically and my body was failing to react. I wanted to move, but I couldn't.

I had spent my entire professional life controlling an animal ten times my body weight, an animal that could bolt at any given moment. I had trained it to gallop at 35 mph, jump six feet in the air, clear a ditch, and control my balance through it all. And I had done this mostly with my legs.

Those same legs now gave way. I collapsed to the ground. I tried to muster the strength to lift myself up, but I couldn't. Instead, I sat down on the kerb. Joanna sat next to me in silence, neither of us having the courage to speak. After what seemed like an endless stretch of time, she supported me back to the house. I lay on the bed, where my body started to vibrate, such was the shock to my brain as it realized just how incapable it was of facilitating the faculties of my body. I needed to be sedated by a doctor, and worse, I was confined to my bed for two weeks.

But I wasn't going to be scared off. I wanted to push everything to the limit. And beyond. The minute I could get myself up and out, I did. But this time, I wanted to go alone.

My plan was to slip out unnoticed when Joanna was out of sight. So I picked up my sticks and had barely taken one step forward when my legs buckled and I fell to the floor. It took some serious manoeuvring to pick myself up off the ground and on to my bed, and when I finally did, I was gasping from the exhaustion of it all.

I told nobody what had happened, but if anything was going to shake me up, this did. I was clutching to the belief that I could do whatever I wanted, but this was unforgiving stuff. And suddenly I realized – in cold, hard fact – exactly where I stood in this game I was playing. Who was I fooling? *I was a twenty-eight-year-old brain-damaged man who couldn't walk.*

But there is soul in my story. And it wouldn't give up. You do not give in to fear. If I was fooling myself, I would continue to fool myself. It was better than giving up. I was convinced that the next time I would do it.

Two days later, I told Joanna that I wanted to rest, and when she went outside my room and closed the door behind her, I got myself out of bed. I picked up my two sticks and with them, I picked up all the broken-down will I could muster. I slipped out the back door and into the fresh morning air. And with nobody holding my hands, I did it.

Ten weeks after my operation, fifteen months ahead of schedule, I did it.

I trumped FEAR with BELIEF.

I walked fifty yards. Fifty yards – just me and my sticks.

Then I stopped. Not from fear, but from hope. And I looked up at that perfectly blue sky and, for the first time since my accident, I saw light. That rainbow? It was mine.

My next step was getting off my medication. Strange as it might sound, instead of making me feel well, the drugs were making me feel unwell. They were shutting off my brain at a time when I wanted nothing more than to feel. It almost didn't matter what I felt, I just wanted to feel. With the medication I felt nothing.

But it was a risk, and for a man who has never gambled in his life, to gamble *with* his life was a biggie. The magnitude of this wasn't lost on me. Because, of course, I was flipping a coin and there were only two outcomes. If I didn't get it right, I'd get it wrong.

This was what the doctors feared, what they had hinted at in shades of grey during almost every conversation they had had with me since I woke from my coma. This is the stuff that they don't ever tell you directly, instead they imply it with their carefully chosen words, their unknowingly patronizing body language. But read between the lines, and there it is – tucked within the innocuous-sounding medical parlance – a not-so-subtle, not-so-innocuous message. When Professor John Miles sat perched on the edge of my bed and spoke to me about the potential ramifications of the trauma to my brain, I had no reaction. None whatsoever. And this is what scared him because it reeked of one of those 'conditions' that medical people seem so inclined to slot you into – the ones with uncertain outcomes that invariably end in the word 'syndrome' or 'disorder'. What they feared I had was called Post-traumatic Stress Disorder and my failure to react fit its medical definition like a glove.

Within a few days of my first conversation with Professor Miles, I was handed a Patient Information Leaflet (PIL) to 'read and understand'. There, on the rectangular, glossy two-pager, scattered between the helplines and support emails and messages of reassurance, was a definition that I have to admit took me by surprise. This is what it said: 'Post-traumatic Stress Disorder is an anxiety disorder characterized by an absence of emotion, an avoidance of thoughts and feelings from the discussions of trauma, until one day one becomes so flooded by emotion that one cannot shut it off.' It was remarkable, the coincidence, because, on the surface, my asymptomatic behaviour was precisely symptomatic of the disorder.

And so this is what the best minds of the departments of Neurology and Neurosurgery at The Walton Centre were convinced I had; based on documented information, I was more or less a textbook example of it. The danger, of course, according to them, lay in what would eventually – but inevitably – happen to me; that it was just a matter of time before reality would set in, and then, I would snap. It wasn't disclosed to me until much later, but in the days following my surgery, and to a fairly high degree of probability, I was considered a legitimate suicide risk. This also explained the growing collection of bottles that I found accumulating in a neat little cluster on my bedside table. Drugs, it appeared, were the only antidote to taking my own life.

They thought they had me pegged. They couldn't have been more wrong.

The fact is that nobody fully understands the inner workings of the brain – how it reacts, why it reacts the way it does, why this is so different from person to person. We can at best generalize, but beyond that, we can only guess. We haven't advanced that far medically – the brain is too complicated. Our cognitive abilities, even as a single species, are too varied. So they didn't know, they *couldn't* know, how my brain was wired. They didn't know what it felt like to be me, what it felt like to wake up every morning when the life that you thought you lived is not the life that you are living any more. So I didn't want to commit suicide, I wanted to live! And I didn't just want to live, I wanted to do everything and more than I had done before my accident. The fact is – and I probably wouldn't have dared to say this to Professor John Miles – that medical professionals, in as much as they are qualified to know these things, they *don't* know, they can *never* fully know, because the brain is one of evolution's proudest creations. It is marvellous, it is mysterious. And it is a complete minefield.

Then there was that little matter of odds. Everybody wanted to know my odds. Everybody wanted *me* to know my odds. I had little interest in the odds. I belonged to a sport that was defined by odds. The very lifeblood of horse racing rests in the odds. So I knew, more than anyone, just how misleading they could be. Odds are useful because they are objective indicators that determine the likelihood of something that has not yet happened. We rely on them because they cut through the noise and attempt to quantify an unknowable future outcome with a simple numerical expression that most of us can grasp. But, where they fail – and this can make the difference between life and death – is that they don't take into account the alchemy of what makes each one of us unique. They don't take into account human variability.

It is exactly the same with medical odds. Despite their best intentions, they can camouflage the truth. When

Professor John Miles was talking to me, he was giving me his best advice based on his own past experiences. But he hadn't experienced *me* before. So, as qualified as he was, when it came to me, none of what he said qualified.

He was basing his words on medical odds.

And medical odds deal with standard expected behaviour.

I wasn't standard expected behaviour.

I didn't fit the bell curve. I was hanging off it.

Within a week I had weaned myself off medication.

The third step was getting sensation back in my limbs. This one was a toughie. I'd swear again now, at this point, but I don't swear unless completely necessary, so 'toughie' seems appropriately euphemistic. It was a toughie. But it was a bloody toughie. I didn't know where to start, what actions to take. How does one *will* oneself to feel? It seemed the stuff of fairies and faith, and it frustrated me.

So I would spend hours trying to force sensation into my hands. I would bounce a golf ball and try to catch it between thumb and each finger in turn. It was excruciatingly slow but I had all the time in the world. At the same time, I started walking barefoot to try to force the feeling into my feet. This was necessary. The lack of feeling, the numbness, filled me with emptiness, a lack of hope, and that frightened me. Because it was hope that kept me alive.

I could walk now, by myself, albeit heavily supported by my sticks. I loved the independence of it – to do this simple task by myself, for myself, whenever I felt like it. So I'd kick off my slippers and I'd go for walks as often as I could. Joanna and I had moved by now to our own home,

Oaktree House on Duchess Drive. Rather bizarrely, I had put in the offer on the house a few days before my accident. The owner had been gracious enough to hold everything in place until such time that I was able to proceed with the formalities – I will always remember her for her kindness.

It was here, on Duchess Drive, that I would go on my many walks, using my sticks – my constant companions – for balance. One such time, I was walking down the road from the house, alone, barefoot. I was expecting a couple of friends from London to visit me later that day and I wanted to get my walk in, quickly, before they arrived.

As I was walking, I noticed a car coming down the road, slowing as it saw me. What I didn't know was that my friends were in it – they had lost their way on Duchess Drive and were hoping to flag down someone to ask for directions. They'd seen a man walking and, given the way I looked at the time, it was hardly surprising that they hadn't recognized me from a distance. It was only when they came up close that they realized it was me. My friend Colin rolled down the passenger-side window to shake my hand, then suddenly he drew back with a jerk, retracting the handshake, that same hand going to his mouth involuntarily to cover the audible gasp he had just let out.

My feet were covered in blood.

I hadn't even realized my feet were bleeding.

Apparently I had walked on broken glass.

I hadn't even realized I had walked on broken glass.

Toughie. Bloody toughie. Bloody literally.

It took me five months from my accident to get sensation back in my limbs. Most of the sensation came in the form of pain. But I was glad for it. If there was pain, I knew there was feeling and if there was feeling – something, anything, even pain – there was possibility. And possibility is all I had.

You've heard this before, but it's so true it deserves to be said again – our experiences shape us. As human beings, this is our truth. Our attitudes, our expectations, our ambitions, our hopes, our desires are all the product of our life experiences and our responses to them. This is what makes us. This is what creates the spirit within us.

My childhood shaped me; it made me strong. We led unassuming lives, but the unconditional love of my parents, the support of my siblings, the security of belonging within that simple existence, was rock solid. It moulded me into a person who could take on anything. And despite being picked up and hurled the way I had been, from the very top of the world to the very bottom of the abyss, all within a matter of seconds – my spirit remained intact.

Within a month of getting sensation back in my legs, I was walking without my sticks.

Within a couple of months of that, my brothers got reports of me running around Newmarket Heath.

Within a couple of months of that, I sat on a horse for the first time – just sat on it – felt it underneath me, felt its breath on my hands, its muscles on my legs.

And when I did that, I knew I was going to do everything I'd ever done.

Nobody becomes somebody until the fight within them has begun. I didn't have a fight within me; I had a war.

Big Boys Don't Cry

You're a fish, swimming along happily in the ocean with your friends. Suddenly, you find yourself caught in a fisherman's net. He pulls you out as you gasp for air. You wriggle and you writhe and you think you're going to die.

Then, the fisherman takes a closer look at you. He laughs. 'This one's just a tiddler,' he says. And he throws you back to sea.

But the ocean looks different now. Everything is strange. You try to find your friends, but they've swum far away. You try to catch up to them but you struggle, your body is too weak. You've come out of the net. But you're still on your own. It's a lonely place to be.

People said that one minute I couldn't get up from a chair unsupported, and the next minute I was running around Newmarket Heath like a madman.

But you don't go from not being able to stand unsupported to running around the heath in a minute.

It doesn't take a minute.

It takes many millions of minutes. Of pain and burn and despair.

It takes every last ounce of resolve.

It takes losing hope and then – somehow – finding it again.

It takes blood and sweat and tears.

It takes being prepared to crawl through a tunnel on your hands and knees through darkness and danger without the promise of a way out on the other side.

It takes being willing to lose everything, to gain that one something.

That's how I did it.

But there was one thing they said that couldn't have been more true. They said I was a madman. I *was* a madman. I had to *become* a madman. Because only in my madness lay my escape.

When I had just left the hospital and returned home in those early days, when I was trying to walk with Joanna and my two sticks, when the car came over the ramp and I collapsed – my brain had stopped me, and my body went into shock. Slowly, my brain recovered. I got past that point. My brain didn't stop me any more. My body stopped me. It was an ordeal of a different kind.

Whether eventually getting sensation back in my limbs was from my effort or the will of the universe, I will never know, but I never stopped trying. For weeks, I had been forcing sensation into my feet by walking barefoot. Simultaneously, I had been working the golf ball, finger by finger in each hand, to regain the coordination in my hands. It was a slow game, but the rewards were high. Before I had started, I couldn't move my fingers on demand – for example, if I had wanted to snap my fingers I wouldn't have been able to do it. If I had wanted to raise two fingers – or one or three or four – and keep the others down, I wouldn't have been able to do it. I didn't have the coordination. If you had dipped my finger into boiling water, I wouldn't have felt it. I didn't have the sensation. If you had poked a pin – or a jackhammer – into my feet, I wouldn't have reacted; I had no feeling. I had to work hard, toil for every iota of movement, for every tiny atom of sensation, to win back my fingers, to win back my feet. And slowly – excruciatingly slowly, excruciatingly patiently – I did.

It was the first of many trials.

Walking properly – like normal people – was another manoeuvre that not only tested my physical ability but also demanded a level of mental focus that I could never have anticipated before my accident. Within six months of coming out of the hospital, I was walking everywhere, all the time, like a man on a mission. Sometimes I would walk for five, maybe six hours. I would leave the house in the morning and come back in the afternoon. Just walking the whole time.

When I think back to those early weeks in Newmarket, I feel immense gratitude towards Jim Hogan, who would routinely come all the way down from Wellingborough just to walk with me. Being a former athlete, he used to advise me, train me on posture and technique. I relied on him more than I realized at the time – just having him there made me feel secure at a time when I was at my most insecure. Because even though I was walking, I wasn't walking normally. When normal people walk, their entire body participates in the movement. It is mechanical, automatic, not something you consciously think about. And why would you?

But if you do, if you break down the anatomy of a walk,

you will realize how complex it really is; how so many different parts of the body have to move in tandem to keep balance, to propel motion.

For example, the body's centre of gravity lies at the hips – this is the seat of balance. So when the right leg moves forward, the right hip is rotated forward as well. This motion is then transmitted, seamlessly, to the shoulders through the spine. The movement of the shoulders mirrors that of the hip to keep the system in balance. So left shoulder moves with right leg, and right shoulder moves with left leg – they move in sync, rising and falling in harmony.

So really, walking is a highly coordinated effort in which the feet and hips and spine and shoulders all participate to enable the body to move. And all of this, of course, is controlled by that most powerful organ in the body, the almighty brain.

Which, in my case, had been tampered with by hooves and scalpels.

So, my body wasn't moving like that. My feet were being thrust forward – right leg, left leg, right leg – but from the waist up, I was as stiff as a statue. As if the two halves of my body had been placed together hurriedly, without so much as a thought for how they would work as a whole.

I had to teach myself. I had to train my brain. Can you think of such madness?

I'd like to offer a thought. Or two thoughts, actually:

Ever wondered what it's like to wake up to a new body? To go from 10.5 stone of lean muscle to under 6 stone of skin and bone? I can tell you. It's torture.

Ever wondered what it's like to try to regain your old body? To go from under 6 stone of skin and bone back to 10.5 stone of lean muscle? I can tell you. It's greater torture.

I started going to the gym: my baptism of fire. I needed to try to coordinate my body, to strengthen my limbs. The doctors had recommended steering clear of the gym for a year post-surgery. I was there in under six months. As soon as I could walk, I was in the gym. I got there walking. But of course.

The first time I sat on a rowing machine and pulled the handlebar, the full extent of my frailty stood starkly revealed. Within three strokes, I was out of breath. There was a middle-aged man on the machine next to me, rowing effortlessly, breathing effortlessly. And there I was, a former competitive sportsperson, gasping for air, fighting with myself, fighting every urge to give up, to stop. It was humbling. It was heartbreaking. It was unforgiving. I was a grown man and I wanted to cry.

But I didn't cry. I didn't stop. It never happened.

I started using the cross trainer, only I had 3-kilogram weights in each hand that I used as a proxy for the bars. I was huffing and puffing, my legs on fire, my eyes shut, my face distorted with pain. People stared at me and shook their heads, half in wonder, half in disapproval.

I was operating on reserve fuel to begin with and, even from that precious little, I kept going until I had sucked out every last drop. I was struggling to breathe, struggling to hold the weights in my hands, struggling to take another step forward.

People noticed. I didn't know which was worse, those sideways glances of incredulity or the other ones, the ones of pity. Once, I overheard two men whispering between themselves. '*That* man should be in a wheelchair,' one man told the other, covertly looking in my direction as he faced the mirror, pumping iron. But I couldn't blame him. I couldn't blame any of them, really. They were right. I was a weak, frail, feeble man, attempting to do something I obviously couldn't.

I pretended they didn't exist. I didn't stop. It never happened.

I became obsessive about my weight. Paul Eddery, former top flat jockey and my good friend, had these precise scales that measured weight down to the last gram. I borrowed them so I could keep an exact account of the progress I was making. Every day I would try to measure my strength in muscle, analyse the improvement. There wasn't any for weeks and weeks. Nothing. Not a single upward tick in that arrow. The disillusionment tore at my heart.

But if that arrow was stubborn, I was more stubborn. I kept going. I found I couldn't sit still any more. I had a constant urge, a tic, a longing to move. I was *desperate* to move. No sooner was I walking than I made myself jog. No sooner was I jogging than I made myself run.

I used to be a good runner as a kid. I wanted that back. I craved the things that made me feel like myself, that made me feel normal. The curious thing is that normal people, who do normal things, do not consider 'normal' as sufficient. But at the point that I was in my life, 'normal' was a huge achievement.

But even that, even an attempt at being normal, was a battle.

One time, I was running up the equitrack at Warren Hill in Newmarket. It's only a short uphill path, about half a mile long, something an averagely fit person could manage with ease. I had been running for a few weeks now – in my head, I had conquered it.

Ah, but no.

Mirage, mirage, an elaborate mirage.

It turned out that I hadn't conquered it. Apparently, I hadn't even arrived on the battlefield.

When I was about halfway up, with no warning at all, my legs crumbled beneath me. My muscles had spasmed in both legs simultaneously, sending shooting pains from my thighs down to my calves. I felt like someone had shot me in both knees. At exactly the same time, my lungs gave way. I ran out of energy.

Around me green grass merged into blue sky and everything started to spin. I saw lights flash across my eyes, and I thought, I'm going to black out. I slumped to the ground with an agonized, animalistic howl.

While the cramp worked its way through my system, I tried to breathe to keep my consciousness. I curled up into a ball, head down, mouth open, trying to force the air into my lungs. But I couldn't even do that properly. I was fighting and losing. My lungs were on fire. It felt like being pulled off the ventilator. Like my life support was being cut off. I had been here before, in this place with no air. It terrified me then. It terrified me now.

After a few minutes, the cramps passed, my breathing stabilized. But those few minutes felt like I had been there all night, such was the intensity of my anguish.

I looked around me. The first light of dawn was

appearing over the horizon. The birds had just started to chirp. Soon, the strings of horses would come into sight as the vast expanse of Newmarket Heath came to life.

Everything was as it should be.

Except for me.

I was broken.

I never spoke to anyone about this incident.

Never, ever. Not even to Joanna.

It never happened.

Nor the next one, a couple of months later.

That never happened either.

I was out on the heath running, again. It was so early, there was hardly any light; Warren Hill lay swathed in fog. I liked to run in the dark, in the quiet, when no one could see me. I was running well, I thought, my breathing was deep and controlled, and I felt good. Despite the mist, it was unseasonably warm, with just that slight early morning bite to the air. I could feel the dew kick up above my socks as I ran, and I was glad I could feel that familiar wetness. I was just glad I could feel.

All of a sudden, I betrayed myself. I stalled. I doubled over, hands on my thighs, wheezing and gasping for air. Just like that, without warning. One minute, I was running well; the next minute I was immobilized. As if a plug had been pulled, the machine sputtered and then fizzled out of life. I couldn't take one more step forward. The lights were flashing red and frantic. It burned to breathe.

Then, out of nowhere, Sir Henry Cecil rode up alongside me on his big white hack. He pretended that watching a former athlete fight for air at 7 a.m. on Newmarket Heath as the tears rolled down his face was the most normal thing in the world. 'Declan, come and have some breakfast,' was all he said. Gratefully, I did.

It taunts you. Life taunts you. It pulls the carpet from under your feet and when you fall down, it laughs. I could hear it all around me, that shrill, derisive cackle. I plugged my ears. I blocked it out. I had to.

Indeed, when I think about momentum-breakers – incidents such as the ones I had, that stop you in your tracks and slap you in the face with a razor-edged dose of reality – I had so many of them, I have lost count. But I would get back up and I would beat the resolve. I would believe.

It was the only way forward. I was a jump jockey. Obstacles didn't stop me; I cleared them. You *have* to clear them. You *have* to leave them behind. They didn't happen.

But subtly, something else did.

People said I changed after my accident. My friends, my brothers, Joanna – they all said that I wasn't the same person. That I became self-involved, impatient, selfish. As if the whole thing took over and took me over. I became a man consumed by myself, by my recovery, by trying to get back what had been taken from me. So they all shook their heads and said I'd changed.

And now I want to respond.

See, it's like this:

Miracles don't exist.

But you fool yourself into believing they do.

You fool yourself because you have to fool yourself.

You start to believe in miracles.

You start to be your own fool.

So had I changed?

Yes. I had changed.

I want to scream this from the rooftops.

I had changed. Because I *had* to change.

The Christmas after Haydock Park, Joanna took me to Barbados. She took me there to relax, but it was difficult to sit still. I had to keep going. The surroundings were different but the goal hadn't changed. And neither had my resolve.

I started doing pistol squats on the beach every morning to build muscle in my legs, to get my balance, to get stronger. And for the first time since the accident, I started to feel pain in my muscles. REAL pain.

Strange as it might seem, before this point, I never had the luxury of reaching the stage where I had muscle pain, because I would always run out of energy first. Now that I had built up a minimal threshold of fuel inside me, I could focus on building my strength. The more I built up my energy, the more I could challenge the muscles in my body. And I welcomed the pain. The searing pain.

My thighs – which you desperately need for riding – were so weak, I didn't know where to begin. The funny thing is I had never consciously thought about my thighs before. I had ridden ever since I was four years old; it was second nature to me. I suppose the very act of riding strengthened my thighs over the years, so the more I rode, the more powerful they became. I had never needed to do things to make them stronger. It felt odd to think about them like this for the first time, about strategies and techniques to build these muscles that I had always taken for granted.

So I did pistol squats. When I started, I had no strength in my core. The first few times, I fell flat on my face, on the sand. I picked myself back up. I brushed the sand off my face. And I replaced it with salt. Then I poured it, the salt, over my cut, wounded pride; crystal by crystal, grain by grain. And I tried again.

For days, I couldn't do more than one or two in each leg. Soon, I was doing ten in each leg. And soon after that, I was doing fifty. I have never felt such burn in my thighs in my life. But I was delirious with joy. The burn meant it was working. I was getting there. Slowly but surely, I was getting there.

There were other things I did – bizarre exercises that nobody had showed me how to do. Crazy stuff, the kind of movements that test the limits of human capability. I simply made them up in my head and then put them to the test – the more difficult, the better. I would go until I was near collapse. Till the sweat was dripping off me. Till I was positively nauseous.

When most people cross the lactate threshold, they know it – it's your body's way of telling you to stop what you're doing. I was so weak. *So* weak. So completely weak. But, I had no threshold. I *wanted* to stimulate pain, discomfort, agony. The more, the better; bring it on. I welcomed it even as I battled it.

But *God*, the pain of it; the *sheer, agonizing* pain of it.

When I returned from Barbados, I went back to being a regular fixture at the gym. Within weeks of my return – ten months post-op – I was routinely breaking my own personal record on the rowing machine; my best time was 5:40.1 for a 2,000m sprint.

How my heart and lungs didn't give up, I don't know.

Desperation – because it *was* that – had burned a hole into my psyche. It was immune to giving in to the impulses of normal people. Fatigue? Exhaustion? I didn't feel it. Instead, I was driven by some sort of unstoppable functional pragmatism – if I kept going, I would get there, I would somehow reverse time. I was convinced of this. So I kept going.

A sane person would have stopped. Something inside them, a safety mechanism, would have stopped them. You *know* when you need to stop.

I didn't have one. A safety mechanism. I had crossed that Rubicon a long time ago.

So, had I changed? Had I become a different person?

It's not a difficult question to answer.

There is a thin line between sanity and insanity. In the eighteen months following my accident, I travelled along, walked along, ran along this line many, many times.

You don't *do* the kinds of things I did in your normal outfit, in your day-to-day existence. To do them, you have to step out of yourself and take your mind to a different place. You have to cross the pain threshold many times over. You have to laugh at despair. You have to battle demons. You have to lie to yourself and you have to believe your own lies. You *have* to become someone else.

There is no shame in saying it. Yes, I changed. Yes, I became someone else. I became mad. I became insane. I became whoever I needed to become. I did whatever I needed to do – *to survive*.

There comes a time in everyone's life where one has to choose between the world and oneself, take sides. I chose myself. If that is a crime, I accept it. Guilty as charged.

As I write this book, I often wonder what I would do if this happened to me now, having gone through it before. If I had to do it over, if I had to go back and start from zero, would I be able to? Would I be capable?

This isn't a hard question to answer either.

My answer is unequivocal: I would fail.

I believe something. Something that is as beautiful as it is melancholic. I believe that a horse gives its absolute, all-out best only once in its lifetime.

Only once. The next time around, and the times after that, it can come close, very close, but it cannot repeat that singular, stratospheric performance of a lifetime ever again. I had seen this of many horses in my time as a jockey and it held true, even for the best and biggest names of equine fame.

It is oftentimes the same with people. It was the same with me. Because when you go that far, it tears something inside of you.

I won my battle, eventually. I came out on the other side. But there is, and will always be, a small gaping wound in my heart that can't ever be filled. I live with this in peace. Not in the hope that my wound will someday be healed, but with the knowledge that it will not cut any deeper. This was the cost of survival, the price I had to pay in exchange for my life. I don't fight it. I accept it.

But then, at that time, nothing was impossible. *Everything* was possible. That's how my mind worked.

The truth is, I was lucky. I had no benefit of perspective. I had no holes in my heart.

I had no idea how difficult it was going to be. I wasn't tarnished by past experiences; I wasn't scarred by past wounds. I locked myself up inside my own head. And I built a house. Yes, I built myself a house and I lived inside it for eighteen months. That's how I did it. I was living in a house of cards.

As soon as I first started walking, I knew more or less how far I could walk without struggling. Every time I did this, my mind would identify a point, a goalpost. And I would convince myself that if I could just get to that point – *if I could just get there* – tomorrow I would be able to go past it. When tomorrow came, and I did it again, my mind had already gone past where it was the day before.

This is what kept me going. These mind games.

Many times I'd find my body struggling to reach my goal, but somehow my mind would take it there – with the promise of tomorrow. 'If I can get there today, tomorrow I will get beyond it.' But the corollary was equally true. And far more menacing. 'If I *don't* get there today, tomorrow will not happen.'

What worked? Reward or threat? Possibility or fear?

I will never know. But you play these games. You trick the mind.

I used the same strategy in the gym with the rowing machine and the cross trainer. A rowing machine has ten resistance levels. Every day, I would convince myself, 'Today, if I can do ten minutes on level one, tomorrow I will do eleven minutes on level two.' I kept shifting the goalpost. Of course, the obvious – and very real – danger was that at any point in time I could have failed to reach my goalpost. And then, I would have realized the sad truth about imaginary friends – they don't exist. Tomorrow may never have come.

I am convinced that if I had been defeated at any one of the times that I had challenged myself, if I had lost even one of those challenges, if I had allowed myself to, I would have failed to reclaim myself. My mind would have given up. It really was that precariously balanced – my confidence, my belief, my faith. Because I had so little to fall back on.

This is why the incident on Warren Hill broke me the way it did. It wasn't my mind that had given up on me, it was my body. It was physical. And my mind had no choice but to concede. But your body is a cage that you are held captive in. So, you dust yourself off and you go at it again. It's the mind. Pure mind. And if your mind is strong enough, you can *will* yourself to break free.

Then, there was the matter of the clock. And the clock inside me never stopped ticking. Time took on a certain gravitas. Targets became numerical. I became recklessly competitive. Time had slipped through the gaps of my clenched fist. I was going to get it back, and not only was I going to get it back, I was going to get it back with interest. If I had lost Time in four years, I wanted to regain it in two. Or less.

I set myself a wildly aggressive deadline to go back to being the person I was – at least physically. For all practical purposes, my timeframe was totally, completely unrealistic. I admit it openly. It wasn't surprising therefore that people judged me. Everyone thought I did it too soon, too suddenly; that everything was too quick, just all too much.

It's true. It really was.

What I did, the whole thing, was nothing short of an act of lunacy. Doctors hadn't expected me to walk again. I got myself passed fit to ride. People warned me not to go near a gym for a year. In a year and a half, I was ready to race. You have to be crazy to even think you can do it. And then to convince yourself that you can. And then to actually do it eighteen months after you have read your own obituary?

No wonder everyone thought I was on the brink of madness.

The truth is that I would never, never, EVER advise anybody to put their body through what I put mine through, in the timeframe that I did. Nobody with a sane mind would recognize how weak their body was, how compromised, and then proceed to fight it and abuse it the way I did. I admit it may have been reckless. I admit it could have been dangerous. But the bottom line was that it got me to where I needed to get to.

It took discipline. Complete and total discipline. Discipline like I never thought I had. They talk about things like this in the army. Discipline and endurance and mind over matter. But they are still different beasts than the ones I was grappling with. Sure, army drills are gruelling, sheer tests of sufferance and strength – tough, tough stuff. But before you even put yourself out there, you've got the resources, the raw material in place. You join the army as an able-bodied person. And if you fail, it's the mind that has failed you.

With me, at that point, *all* I had was my mind. My body didn't exist. And many, many times my mind tried to carry my body, but my body wouldn't obey. Those were the times when I wondered how long I could carry on. How long I could keep fooling myself. How long before the cookie crumbled. Those were the times I (never) cried.

I know with certainty that there were many who were convinced that I had medical-grade post-traumatic stress, that I wasn't accepting what had happened to me. So they walked on by. But many feared an unhappy ending: 'Poor, deluded fellow, he's pushing himself so hard, thinking it's possible. When he realizes it is impossible, he will break.'

But they didn't understand me at all. It wasn't that I hadn't accepted what had happened to me. That seemed almost irrelevant somehow. I just wanted to go back to being me. So through this whole ordeal, it was never a case of feeling sorry for myself. Frankly, I didn't have the time to feel sorry for myself; I was too busy, too desperate trying to *become* myself. Whoever that was.

And somehow, eventually, I did. Apparently, in a minute.

This chapter in the book was written last. For a reason. It wasn't going to exist.

When Ami started writing my story, she told me it would hurt. I pretended I hadn't heard those words, and we moved on. She demanded honesty. I gave her honesty. But I wasn't ready for this.

I had never told anybody this before.

Not anyone. Not ever.

Let alone other people, I hadn't even allowed *myself* access to my thoughts or feelings. Introspection? Self-reflection? No chance.

I just never wanted to go there again, such was the intensity of pain I felt, even from the recall.

I had always been a strong man. A confident man. And I had been brought to my knees. I didn't want anyone to know this. You never want to be seen as being that weak, as being that vulnerable. Because if you declare that, in the situation I was in, if you show your cards to anyone, *even to yourself*, you give up hope. And when you have nothing in your camp but hope, it's not something you can afford to abandon.

It is a long minute that takes you from being incapacitated to functioning to riding in a race. It's a minute that lasts eighteen gruelling months. And during these eighteen months, you never rest. Not once. You can't.

There's a three-way war raging inside of you.

Your body wages war with your mind.

Your mind wages war with your spirit.

Your body is the weakest, first to give up.

Your mind is stronger; it can take your body to places your body didn't think it could go. But it is also fickle. Once it knows fear, it backs off. It is superstitious. If things go wrong, it cowers.

It is the spirit within you that is unbreakable.

The spirit unshackles fear, it lifts the mind, it empowers it.

The spirit has *got* to win.

Otherwise, everything is lost.

My spirit was my lifeline. It fuelled my mind. In turn, my mind fuelled my body. Everything was built on sand. Had my spirit broken, had I allowed it to break, my world would have collapsed – it was that brittle. One admission of defeat, one failure, one loss could have been all it took. My house of cards, so painstakingly built, would have come crashing down. With me trapped beneath the rubble.

That's why I've never spoken about this. Because I know how close I got; how narrowly I escaped.

I still don't understand how she manages to get it out of me. But no sooner am I out of the driveway, following our meeting, than I call her from the car to tell her I am not prepared to put this in the book.

'Pretend it didn't happen,' I say.

'Which part? What you put yourself through or our conversation?' she asks.

'Both,' I say.

'OK,' she says, 'it's your book.'

An hour later, I call her back. Talking about it has opened up old wounds. I feel the pain – dull and heavy and as overpowering as the pain that my body went through at the time. The pain is mental, but I feel it in my body like it is physical, like I am reliving the whole thing. I can hardly bear it.

When she answers the phone, she's cautious. She knows by now when I am fragile.

'Go on, put it in,' I say. 'What the hell, you're writing my book, what's to hide.'

'Are you sure?' she asks.

'Yes, I'm sure,' I say.

'It's the right call,' she says, 'and a brave one.'

'But there's one thing,' I say. 'You know all those times I said I cried? That bit never happened.'

'Of course not,' she says. 'Big boys don't cry.'

Entrapment

I am a spectator in a theatre of terror.

My mind is a concert hall, vast and venerable, a fitting venue with its air of grandeur, its perfect proportions. There is a stage in the very centre, the bold platform where the drama will be enacted. The cast of characters is varied, they come and they go, but I know them all intimately from different periods of my life.

And I am my only audience.

But when the performance begins, I realize it frightens me. I don't like it. I want to leave but I am shackled to my chair by invisible chains. I can't get up. This happens many times. Many, many times. Sometimes I break free, run to the sides of the room and frantically push against the walls to find a way out. But there is none. On the stage, the ghoulish scenes continue to unfold as if oblivious to the terror of its audience. Sometimes the cast jumps off the stage, runs up the aisle and stands beside me. I am terrified they will touch me, but they never do. I run away from them, but they run with me. Following me, like my shadow.

Slowly, like a python, the panic uncoils and begins its

slow rise inside my throat. Until it swallows me, whole. And I am trapped.

Inside the belly of the beast.

Inside my own terror.

The terror of a nightmare when you are asleep does not compare to the terror of a nightmare when you are awake.

It was a little over seven months after my surgery. I could walk, I could run. To the outside world, I had recovered. I was normal. Inside my mind, the streets were burning.

One day, Barney Curley and his son Charlie paid me a visit, unannounced. I was alone in the kitchen in Oaktree House when they arrived. Barney looked dishevelled, dressed in trousers with braces, a crumpled shirt, no hat. He took a long puff on his slender cigarette and exhaled – slowly, deliberately, blowing circles through his lips.

Then, he opened his mouth and began to speak, but the words were muffled by the fumes and I couldn't understand them. His soft voice boomed loud, ricocheting off the walls, echoing spookily around the small room. I put my hands to my ears to shut out the noise.

But then the images started flashing before my eyes.

FLASH . . . Barney Curley, ten years younger.

FLASH . . . Barney Curley, in a suit and shoes.

FLASH . . . Barney Curley, in the hat he always wore.

FLASH . . . Barney Curley, on *The Late Late Show*.

FLASH . . . Barney Curley, a ghost of himself.

Even as I was looking at him, I wasn't seeing him as he was standing before me; I was seeing him as he was inside my head. He was on TV, sitting on a chair, legs crossed, shoes polished, speaking to Gay Byrne. I was seventeen years old and I was watching him with my mother.

But he wasn't on TV, he was in my kitchen. He wasn't sitting, he was standing. He wasn't wearing the hat he always wore. He wasn't speaking to Gay Byrne. He was speaking to me, but I couldn't make sense of what he was saying. I was twenty-eight years old. My mother wasn't with me.

I shut my eyes in confusion. When I opened them, the plume of his cigarette smoke was wafting upwards in a long, slow spiral. Inside me, the panic took life, uncoiling, rising. An angry serpent. Then, through the curtain of smoke that hung between us, I saw Barney's lips curl, in slow motion, into his characteristic half-smile.

In my head I heard him snarl.

My blood curdled.

And I ran.

I ran away.

Like a frightened child, I bolted past the bewildered father and son and, shaking with fear, I ran out of my own house and down the road. I ran as fast as I could, hysterical with emotion. Behind me, Barney and Charlie jumped into their car and followed.

Frenzied and afraid, I ran down Duchess Drive, trying to break free of myself. I ran and I ran, as far as my weakened legs would carry me. Then, when I could run no more, I ducked into a shed, behind some bushes at the side of the road.

Losing sight of me, Barney drove past where I was hiding, then parked the car and got out with Charlie. They called my name, frantic with worry, desperately trying to find me. But I lay crouched behind the bushes, silently watching them, failing to make sense of the chaos in my mind.

I stayed there for a long time, long after Barney and Charlie had given up, sat back in their car and driven away.

There were riots inside my head.

I had tried to run away from them.

Run and hide.

But I found that they were running with me.

Following me, like my shadow.

There was nowhere to hide.

It was complete entrapment.

Heartbeat

They say you can run, but you can't hide.

I didn't listen. I ran and I hid.

But what do you do, when there's no place left to hide?

When you realize that there's no one left to save you from yourself?

When that noise inside your head becomes unbearable?

Where do you go, then?

I started going regularly to Mark Tompkins's Newmarket stables. I found it therapeutic to be around horses. The first several times I went, I didn't ride. I didn't even move.

I sat on a horse. Just sat on it.

I mounted it, my seat in the middle of the saddle, my knees not gripping, my legs simply lying against the body of the animal, feeling the tightness of skin over hard muscle, velvet over steel.

I sat on a horse and I felt it: the warmth of its body against mine, each muscle independently alive, twitching, responding, to my own tentative touch.

I sat on a horse and I watched it: its breath blowing from flared nostrils, clouds of mist in the cold, early morning air.

I sat on a horse and I smelled it: the dusty, earthy, grainy richness of it.

I sat on a horse and I heard it: exhaling, nickering softly, that low, reverberating rumble from deep inside its throat.

I sat on a horse and I touched it: my hands caressing the convex contour of its crest, the rising arch of its back.

I sat on a horse and I sensed it: the physical consciousness of the sheer size of the beast, of the power underneath me.

I sat on a horse. Just sat on it.

Nobody could understand why I did this, why I went every day just to sit on a horse. But when I did it, just for those few minutes, I borrowed my freedom. And I wanted to hold on to that feeling, for ever.

For over a year after my accident, I was afraid of being alive. I had managed to repair my physical self. I had fought to do this, to claw myself back. But my mind remained in tumult.

It is impossible to explain the complete, utter state of confusion when you are sometimes a child with adult thoughts and sometimes an adult with child thoughts. When a chunk of your life has gone. When what's come back is disordered and incongruent, like pieces from the wrong jigsaw puzzle that, no matter how hard you try, will never slot in. When everybody around you believes you have recovered because that's what you've tried so hard to make them believe. And yet *you* know you haven't. And perhaps you never will.

Still waters run deep. Belief ebbs away and you capsize under your own weight. Everything is an illusion. You learn not to trust your own eyes. Or your own mind. Often they contradict each other and you are never quite sure which is right and which is wrong, what's real and what's a cruel, cruel trick.

The incident with Barney Curley was not an isolated one. I had been having hallucinations ever since I had woken up from my coma, mostly when I was asleep, but many times when I was awake, when I was fully conscious. I had been warned about these 'dreams' early on by the doctors. I had been told that they would vary in their intensity and their severity, and then gradually disappear with time. But I hadn't realized how slowly time would crawl. Or how frightening it would all be.

I had long stopped my medication, so I knew it wasn't the drugs that were causing the confusion. I had nothing to blame. To lash out at.

It was me.

And I don't think anybody will ever understand how far I wanted to get away from being me – the me that I was in the year after my accident. Because being me was chaotic. I had no idea what was going on inside my head. I had no control over it – the disorder, the commotion. When my mind warped, it terrified me, because it swung me around like a pendulum, between my childhood at one end and my adulthood at the other. And often I found myself stuck in the scary spectrum of my own past, a grown man reduced to a helpless, whimpering child. And like a child I had tried to run, but like an adult I had learnt that it was a race I would never win. You cannot outrun what is running inside you. You cannot escape yourself.

I found it so difficult to be around people. They didn't understand me. *I* didn't understand me. That's why I went to Mark Tompkins's stables. There was no judgement with a horse.

It is amazing what a horse can do for your soul.

I had always felt this, even as a little boy with my ponies, but I had never needed to rely on it in the way I did now. Horses empower humans and there is a reason for it. Because the horse is a prey animal, because survival instinct defines its existence, its response mechanism is immediate. So even at the quietest moments, it is intensely aware of its surroundings, acutely able to understand human behaviour, and reflect it. This endows the horse with an incredible gift; it enables the animal to be a perfect mirror for our feelings. I found that horses accepted me for who I was, not who I was supposed to be. In doing so, it granted me the validation I was so desperately seeking. It allowed me to get in touch with my own feelings, to accept myself for who I was, and not who I was supposed to be.

I have always believed that horses are wonderful with people who have physical or mental disabilities because they are able to bring out the human element that often lies dormant within us. Horses are intelligent and honest, and their innate ability to read human emotion is perhaps what lies at the core of the man–horse relationship. Patient and gentle, they offer their insight and empathy, responding naturally to emotional issues, offering their companionship, their silent support. And in doing so, they enable you to win back trust in yourself.

I always believed this, but I never imagined I would put it to the test. In my conceit, I never imagined I would need to. But I did. A disability can come in many forms. I may have been physically able, but my mind was badly – so badly – disabled.

At some point in our lives, we come to the inevitable conclusion that so much of what we do and how we live is influenced by factors outside our control. For most people, the only thing we can control with certainty is our own mind. Our thoughts are based on our own decisions to have them. This wasn't the case with me. I had no control over my thoughts, the voices in my head – I was hostage to myself. After each episode of the kind that I had with Barney Curley, I had to fight not to feel defeated, to get back up again and keep going.

To this day, Barney has never made mention of what transpired at Oaktree House that morning, and for that I will be grateful to him for ever, because nobody understands the courage that's required to face the day after. You find yourself alone. And no matter how strong you are, how brave, how fearless – it debilitates you, the isolation. The definitive, unmistakeable realization that you are well and truly alone. You try to drive it away – this sense of exile – in whatever way you can. For many months, I had tried to fight it myself, on my own. But in the end, I had failed. The only match for a wayward mind is a restful soul. And so I turned to horses to help repair my soul.

It was cathartic.

For close to six months after I started going to Mark Tompkins's stables, I didn't actually ride horses. I sat on them, I walked with them, I watched them. Sometimes I just stood on the dirt road on the outside of the long, white, continuous fence that demarcates the border of

Newmarket Heath, and took it all in from a distance.

Watching . . . as the massive strings of horses walked past me in all their glory and grace. Listening . . . to the beat of Newmarket Heath, thrumming with the sound of twelve thousand hooves pounding against the earth.

Other times, I would get up close to a horse, my gaze searching the eyes of the animal, searching for a connection.

I was anxious.

I was trapped.

I was losing my mind.

Horses became my meditation.

Horses became my escape.

Horses became my sanity.

A year after my accident, almost to the day, I knew it was time.

It was springtime in England, the daffodils bloomed like countless shining suns, the wind felt almost warm. Newmarket Heath was resplendent with the magnificence of galloping horses across its vast and boundless sweep.

And I knew I was ready. I felt it from the inside, that familiar urge, that longing.

It was a deeply poignant moment in my life – the moment that I accepted that I was ready to ride again.

For many weeks, I didn't go any faster than a trot. It gave me comfort, that steady two-beat rhythm, my body following the tempo, rising up and down with every other beat.

Two beats.

Two bodies.

Two souls.

I felt as if we were trotting through time, through this deeply personal, deeply heartbreaking passage in my life. And in doing so, we formed a strong connection, a bond based on this unique form of communication, on this sense of solidarity where I felt, even without words, the horse speaking to me, 'You are not alone. Whatever is happening to you, I am with you. We are in it together.'

I found the experience intensely moving. It returned something to me, something I had naively allowed to be taken away – it gave me back my perspective. Ever since I had woken up from my coma, my mind had been full of clutter. Focusing on my missing past, on my uncertain future. The thing about a horse is that it always knows when you are present with it. It knows if your mind is focused, if it is actively engaged, or if it is somewhere else, a million miles away. The horse made me focus on my present. On my *now.* And I realized something immeasurably profound. I realized how fortunate I was to have a *now.*

This life, this precious life, was mine to lose.

A few weeks after I first started riding, I quickened my pace.

Two beats to three.

From the trot to the canter, that beautiful, flowing gait.

Hind leg, hind leg, fore leg; the most expressive gait of them all.

It is the sequence of footfalls, not the speed, which establishes the canter. So I focused my mind on sound – the beating of his hooves, the whistle of the wind through his ears – as we rode together, our bodies in rhythm. The beat of the canter, calming my mind, quieting the chatter.

For someone who never wanted to be a jockey, it was ironic that being around horses helped me to heal. I took immeasurable comfort in our togetherness.

It drove away the isolation.

It absorbed my mind.

It took me away from the pains of my past.

It gave me balance at a time when I desperately needed it.

It brought me back from the brink, and carried me home.

Fifteen months after my accident, I was ready to gallop.

Two beats to three.

Three beats to four.

My first gallop since Haydock Park.

The horse Mark Tompkins brought out to me was Staunch Friend. He did this deliberately and although I didn't remember it, this horse and I had made history together, winning the Bula Hurdle in December 1993. He was a beautiful-looking horse, a big bay gelding with a tremendously long stride and a striking head carriage, confident and in control.

I saddled him myself, then I mounted him. We walked a bit up the dirt road, then he broke into canter as we approached the heath. But I knew he wanted wings. And I felt ready to fly.

When you are all alone, you cannot abandon yourself. But it helps to have a friend.

That day, when Staunch Friend and I galloped together over the sweeping plains of Newmarket Heath, I felt my mind relax. I listened to the beat of his stride, the four-beat rhythm – hind leg, hind leg, fore leg, fore leg – the gait of the classic racehorse. And as he kicked up the dust with the beating of his hooves, everything felt perfect, in total harmony and complete peace. I felt the familiar addictive thrill of riding a thoroughbred at speed; it filled me with a contentment I hadn't felt for as long as I could remember.

At your finest hour as a rider, it is trust that joins you and your horse. You put your life in that horse; you give it your complete trust.

That's how we rode.

Flying together through the wind, man and beast; of one mind, of one heart.

Dancing, soaring, creating light.

Claiming back my freedom.

And with my freedom came another realization. That morning, on the gallops at Newmarket Heath, I learnt something about myself; something I had always known, but had never been able to explain. I learnt why I was never meant to be a jockey. One of the greatest deceptions we can fall victim to is the myth that relationships need to fit into pre-existing boxes. It simplifies things to believe this. And so we do. Horses and I had always had a strong emotional connection. We understood each other. Horses were my first friends; some of my best friends. And so it seemed – in the eyes of the world – that the obvious way to channel this relationship, to define it, was for me to become a jockey. But how wrong that was.

My experience over the past months had shown that there was a greater purpose to this connection I had with horses. There was a purity to it, a magic that was at its best when left in its innate, primitive state. Interfering with this, the natural order of things, was only corrupting its beauty.

I was born to ride, but I was born to ride free.

When I dismounted that day, the first time after I rode a gallop on Staunch Friend, I walked alongside him, shoulder to shoulder, and he nuzzled against me.

I looked at him, at his eyes. His beautiful, soulful eyes . . . a window into myself.

And he looked back at me.

Like he felt my sadness.

Like he wanted to take it away.

I was overcome then by an intense and inexplicable emotion. It was the moment that I finally accepted myself; the moment that I finally forgave myself. The moment that I accepted I had lost four years, six months and twenty-four days of my life; that I wasn't going to fight it.

Beside me, Staunch Friend was leading me forward gently. And, for the first time since my accident, the clamour inside my head quietened long enough so I could hear them – our two sounds mingled together as one, so much so that I couldn't tell them apart.

His hoofbeat.

My heartbeat.

Why Me?

I fell off a horse and almost died.

Why me?

Nope. That is not the question.

A long time ago, someone told me laughingly I was so lucky that if I fell off a building, I'd fall upwards, not downwards. I never fell off a building, but I did fall off a horse. That in itself makes me lucky. I may have been a good jockey, but I'm no match for gravity. If I'd fallen off a building, *I would be dead*.

If you think I'm being facetious, I'm not. Things fell into place for me in miraculous ways. The stars aligned. What happened to me belonged squarely in the domain of chance, so the stars *had* to align. It was like a game of dominoes. If one fell, the whole thing would have come crashing down. It didn't. There they were, the stars, all lined up in a neat row. Only for me. I believe I had an inordinate amount of luck.

Let me count the ways:

One: I belonged to a sport where two ambulances follow you around all the time. If I'd suffered the same injury doing almost anything else, *I would be dead.*

Two: The paramedic on course knew exactly what to do with me. If he had rested his hand on the wrong side of my head, *I would be dead.*

Three: The ambulance driver was having a good day – the traffic around Liverpool cooperated with him. If he had brought me in to Warrington Hospital four minutes later, *I would be dead.*

Four: The Walton Centre lies within 26 miles of Haydock Park. Had I been on any other racecourse in the country, *I would be dead.*

Five: There happened to be a professor of neurology at The Walton Centre called Professor John Miles. If he had been out of the country instead of out playing golf, perhaps *I would be dead.*

Six: My father had a morbid fear of flying. If he hadn't, *I would be dead.*

Seven: I was a professional athlete, physically at the peak of health and fitness. If I had been of a weaker constitution, *I would be dead.*

Eight: They said, even if I survived, I might never walk. That I would lose my sight, my hearing, my speech. I would exist, but I wouldn't live. *I would be dead.*

I look up at the night sky. And I stand in awe of my luck. The stars haven't lost their twinkle.

Never: Why me?

But: Why me?

That is the question.

A Little Concept of Time

> 'Clocks slay time . . . time is dead as long as it is being clicked off by little wheels; only when the clock stops, does time come to life.'
>
> William Faulkner, *The Sound and The Fury*

3 seconds: to end my career.
4 minutes: to get to the hospital, alive.
3 hours: to survive, for a 50 per cent chance to live.
96 hours: to wake from a coma.
10 weeks: to walk 50 yards, with sticks.
20 weeks: to get sensation back in my limbs.
6 months: to walk unaided.
7 months: to run like a madman around Newmarket Heath.
9 months: to sit on a horse for the first time since the accident.
12 months: to feel the rhythm of a horse in motion.
Tick-tock, tick-tock, tick-tock . . .
It's a race against time. But I'm a race-rider.
18 months: TO RIDE, TO RACE, TO WIN.
Timeless: to find my soul.

Centaur

When I run, I choose the hilliest course I can. I do this with deliberate intent, and I do it not for pleasure but for pain.

Hills are not my enemy; they are simply an obstacle, meant to be overcome. So when I approach a hill, I don't wince, I focus. I shorten my stride, control my breathing, power my leg muscles and I drive forward, my body standing tall.

And I fight those hills. We are capable of infinitely more than we think.

I remember the precise moment in time that I decided I was going to make a comeback. I knew it with a certainty that was almost startling in its clarity.

I was sitting on the sofa in my house in Newmarket one afternoon – about ten months after my accident – flicking through the TV channels. There seemed to be nothing of particular interest to watch, and so I switched the TV off. And right then, in the shiny blackness of the TV screen, I saw the reflection of a painting hanging on the wall behind me. It was by Neil Cawthorne, oil on canvas, and there was something profoundly telling in its simplicity.

Art evokes emotion in me. It moves me from the inside. It always has. And so it was with this piece.

Aptly titled *The Long Way Home*, the painting shows a jockey from the back, leading his horse away from the racetrack. In front of them is the looming majesty of the grandstand at Cheltenham. The horse has fallen three fences from home, and the rider has had to abandon the race and walk away. The air of despondency is apparent in the body language of the rider – his head hangs low, his shoulders weighed down by the weight of something indefinable.

I know so much about all this because the rider in the painting is me.

But now I looked at the piece with different eyes.

I was known in riding circles for my seat in the saddle, for the way I had perfected the classic 'Martini glass' posture – a 45-degree angle between my shoulders and my knees, another 45 degrees between my knees and my hips, a straight line between my knees and ankles, perpendicular to the ground. When I was racing, they said you could balance a ball on my back and it wouldn't fall off. When I wasn't racing, I stood ramrod straight, shoulders back, head held high. It was the body language of a man in control, full of vitality and confidence. It stood in total and complete contrast to my posture in the painting. There was a dejection about me there. A restlessness. A yearning to be where I belonged – *in* the race. But there I was, walking away from the race, taking 'the long way home'.

And now it gnawed at me, this feeling.

It had become deeply familiar, this sensation, in these last few months, so much so that my bones ached with the weight of it. 'Maybe one day you will walk,' the doctors had said. In less than a year, I had beaten the statistics well and clear. I could walk, I could run, I could ride, I had built up my energy, my muscle, my strength. It had not only been a steep slope, but a slippery one, dotted with dangerous 'what ifs?' that had often threatened to shake my resolve. But I had crawled my way up on my hands and knees, yard by yard. And yet, it didn't seem enough. I was restless all the time, searching for something, I knew not what.

What I did know was that this unknown, unspecified longing was growing bigger and bigger and bigger every day and that if left unchecked, it would soon breathe life. And then it would fight me. I had to stop it, before it stopped me.

I stared at that painting for a long, long time, even as afternoon melted into evening and evening into night. The light changed, but the essence didn't.

Sleep evaded me that night, my breathing shallow and laboured as I lay tossing and turning in bed, troubled by the ghosts that hadn't left my side for eighteen months. I longed to rid myself of them. So I could breathe. So I could sleep.

I sat up in bed, my eyes adjusting to the darkness and thought – one last time – about the painting in my living room. And in that instant – that still point in time – I knew I was not going to let destiny decide my future. I was going to create my own destiny.

I was not ready to walk away.

Not yet.

I was going to race again.

And finally, I found sleep.

*

In October 1995, eighteen months after Haydock Park, twelve months to the day I could walk again, I made my comeback. It was a decision made in flagrant disregard of the orders of my doctors and the concern of my friends and family. Perhaps they were right to be worried. Perhaps anyone who had escaped the jaws of death the way I had, had no right to tempt fate. Another fall involving any kind of trauma to my head would kill me. It was that simple. So they warned and cautioned, coaxed and cajoled, reasoned and rationalized, pleaded and prayed.

But they didn't understand.

This was no show of bravado. I wasn't anybody's hero. I was just a man searching for his soul. My decision to ride was driven by one desire and one desire alone – not to be seen as someone dead.

I decided to ride in the Flat V Jump Jockeys Challenge at Chepstow. The announcement of my comeback caused a stir in the racing world; the interest in me, in the race I was going to be riding, in the horse I would choose to ride, was intense. The media went absolutely berserk – everywhere I went, there was press following me around. My phone rang off the hook. There were cameras outside my house. It was pure madness. When I was a ten-year-old kid, pretending to be Eamon in my first-ever pony race, nothing would have seemed more thrilling than the idea of paparazzi. Now, it all seemed misplaced. I couldn't understand it. Why was everyone so interested in me? It was no different now to what it had been then. I was a child then, I was a man now, but I was still the same person – an Irish boy with an overwhelming desire to ride a horse.

As the day approached, the racing world rallied around me. Jockeys called to wish me well; trainers rang, offering their horses. This was a race determined by draw; trainers entered their horses and the clerk of the course picked names to pair horses with jockeys. I was drawn to ride Geoff Lewis's Jibereen, a beautiful bay colt with a proud carriage and a bright-eyed, exuberant demeanour.

A few days before the race, I wanted to have a sit on the horse to get a feel for him, so I went up to Geoff Lewis's yard and rode him in a workout. I realized, coming out of it, that riding Jibereen to win would require a certain judgement of pace – he had a high cruising speed, but it was going to prove essential to sustain it; travelling even one gear higher could prevent him from delivering his best at the winning post. He didn't quite finish the workout the way I would have liked. On the day of the race, this would prove a priceless education.

Once I was booked to ride and word got out that I was going to ride Jibereen, the late Pat Eddery – eleven-time flat-racing champion jockey – called to give me his advice: 'If you ride this horse, you need to break sharp from the gates, and get across to the rail as soon as you can. Then, just get into a rhythm and try and sustain it for as long as you can, so you can get the pick-up you want when you need it.'

It reinforced exactly what I had felt instinctively from my workout on the horse a few days earlier. Now it was down to being able to execute the strategy at a time and a place when it mattered.

Could I?

Was I unsure?

I was unsure. I was completely unsure.

It's so easy to lose trust in yourself.

But I was riding on the belief that you create your own destiny. So if I was unsure, I couldn't show it. Not to the world, not to myself. I remained unfazed. Such was my determination to do everything I had ever done; to be everything I had ever been. It was clear in my mind: if I was a race-rider before, I was going to be a race-rider again.

I couldn't remember any of it but I was going to go out and pretend that I could.

It was the summer of 1989. We were sitting on the bend at Hollywood Park racetrack, Bill Shoemaker and I. He was tiny next to me, barely 1.5 metres tall, but with 8,833 wins from 40,350 rides, he was a force to be reckoned with, a true legend – cream of the crop, best of the best.

On many a day over the course of my time in California, I would sit alone on that bend and watch Bill ride. Just watch him. And wonder how this diminutive figure used to cause a traffic jam and then just kick off the bend at the precise moment – and not a fraction before – he felt his horse could run all the way to the line.

He took to me for some reason; he'd speak to me with an abrasive kindness typical of the most effective mentor–protégé relationships. I suppose he believed in me, in the fact that I could ride.

One day we were sitting together in silence, watching the horses. It was a balmy California afternoon, sun in the sky, breeze in my hair, when suddenly he peered at me and said, 'You goddamned son of a bitch, you probably don't want to be a jockey but if you change your mind, remember this – you've got to ride the race to suit your horse and not your horse to suit the race.'

And I looked at him, eyes wide in wonder, and I marvelled at his brilliance.

I had heard that Bill Shoemaker was born weighing 2.5 pounds, so small that he was not expected to survive the night. He made it, spending his first hours in the world inside a shoebox, which in turn was placed inside the oven to raise his body temperature. But even at his heaviest, he weighed no more than 100 pounds. The miracle of Shoemaker was that even though he was smaller than everybody else, he rode more winners than anybody else and I realized at that point, that in this game we had both chosen to play, the mind and the brain assume a far greater role than any physical trait a man might have.

This was about instinct and intelligence. That's all.

It was a moment of epiphany.

At Chepstow, on the day of the race, the pressure to perform was immense. But I knew that no matter how scarred my body was, no matter how scarred I thought my mind was, I would need to rely more than ever on these twin traits – instinct and intelligence – to see me through. And I knew that I would need them to stay true to Shoemaker's art of race-riding, to be in harmony with my horse, to 'feel' my horse, to mould my body to its rhythm, to be at one with it, to morph into it, to become – for these next crucial minutes of my life – a centaur.

So, when I got on the horse that afternoon, I felt confident and ready. The horse was heavily gambled on, backed down from 7/1 to 3/1 favourite. I was overwhelmed by the good wishes from my colleagues, my friends and family – if they disapproved, they didn't show it; everybody stood around me, supporting me. It was a remarkable show of solidarity and I was glad for it.

And secretly? I needed it.

Because during the countdown to the start – that final countdown – I felt an emotion I had never felt when riding before. It was alien to me and the unnatural nature of it unsettled me. My brother Pat would later tell me that when I got on that horse I had a look on my face that he had never seen before. 'It was determination,' he said confidently.

He was wrong.

It was doubt.

It was a moment – *just only* a moment. But it was there. That big black cloud of doubt moving silently across the sun, threatening to engulf it.

Because what nobody knew was that I had no peripheral vision on my right side. *I was riding partially blind.*

Nobody knew this. But I did. And sometimes, that's enough.

You can fool the world, but how long can you fool yourself? How long can you keep it intact, when it's not? I think it's right until the point when you realize the solemn truth that the joke is actually on you. And that's all it takes, that one second of fear, to shatter everything you have built up.

Ah, the fragility of confidence.

And with that, out they came. The questions. From inside my head.

Like a bucket of sand tipped on its side, the questions – tiny golden granules of doubt – came tumbling out: Could I still ride the way I'd ridden before? Did I remember the fractions? Could I judge the pace? Was it too soon? What the hell was I doing here?

The phantoms – the ones that appeared before me during those black nights at the hospital – came back, with their white distended faces, the sockets with no eyes, the lipless mouths. I felt like I was twelve years old, thrust once again into a situation too grown up for me to handle, and I thought to myself, Are you sure you are ready to be here?

And then, a voice inside my head. It wasn't one of *their* voices; the phantoms, they didn't have voices. It was my own and it was screaming: 'My God, you are going to get exposed.'

But the mind is fickle.

That day, I used this to my advantage. It passed, that moment of doubt, as fleetingly as it had arrived.

I drove out the demons. I had to. I was an elite athlete. There were no limits – they didn't exist.

Faster. Higher. Stronger.

See it, feel it, believe it.

I was good enough then.

I would be good enough now.

The fragility of confidence versus the resilience of trust; the latter endured.

Still, it wasn't over. Given the absence of my right-side peripheral vision, it would have benefited me greatly if I had been drawn near the rail. Unfortunately, I was drawn in stall seven, in the middle of the track – there were seven horses between me and the rail on my right side.

But I wasn't going to let this shake my belief. I heard Pat Eddery's voice in my head – I was thankful for it. Because more than anything, I couldn't afford to let Jibereen feel my fear. Unlike most other sports, where the player or rider interacts with an inanimate object – a racket or a ball or a bike – riding involves working with a sentient being. And even the minutest of movements made by the rider – touches of the rein, shifting of body weight, changes in leg position, even a break in breathing pattern or a quickened heart rate – can be felt, and reacted upon, by the animal. I knew I could not afford for Jibereen to sense any misgivings I had.

Fear was a hill. I was going to fight it.

I narrowed my vision, and got my head into the zone, that state of single-minded immersion that was characteristic of me before any race. Serene and alert and totally in control. So completely focused was my mind that when my colleague Ray Cochrane turned to me at the starting stalls and said, 'Declan, you haven't pulled your goggles down,' I replied, 'Ray, I'm not going to need them.'

I was in the flow.

And when the gates opened I was out like the wind; *poof*, I was gone.

I had one initial goal – to get to the rail. I got there quickly and within ten strides, I was able to find my horse's cruising speed. As I had predicted during my workout, it was a little bit higher than the horses I was racing against. I knew that the most important thing for me was to keep Jibereen in his cruising speed, to sustain that pace for as long as I could.

Ride the race to suit the horse and not the horse to suit the race.

If I did this, I knew that even if a rival swooped in to take me on, I would remain confident in where I was. Exactly this happened at the 2-furlong pole; I stayed my cruising speed and it appeared for a moment that the field was going to swallow me up.

There was a stillness in the air – a state of dynamic equilibrium – as if even a whisper would disturb the balance of the universe. I could feel it – the pent-up anticipation, the excitement, the expectation. Everybody was waiting. Everybody was watching. I could not let this slip.

Then, a furlong and a half to the winning post, at that point on the track at Chepstow where the horses run downhill to the finish, I decided it was time.

Time to let Jibereen go.

Time to feel his speed.

A racehorse in full flight is a thing of beauty; an artist, an enigma. An elite athlete that bursts into life in a bid to perform. *Every minute* at full gallop, a thoroughbred pumps some 1,800 cubic litres of air in and out of its lungs. Its heart beats 250 times – nearly five beats a second – to pump 300 litres of blood around its body, all to achieve that singular goal: speed.

That day, the light shone on Jibereen. He was performing for me, one breath perfectly in time with one stride as he raced towards the finish, the organs in his body working together in exquisite harmony, pumping the oxygen from his lungs to his heart, from his heart to the muscles that powered his spectacular speed.

And I felt it. At that moment, I felt it, like I had felt it my whole life. The spirit of the animal underneath me: the power and the pride, the swiftness and the strength, the majesty and the gentleness and the grace.

I felt my horse.

I was at one with it.

I was a liminal being.
I was CENTAUR.

I let Jibereen go. He quickened half a stride. It was enough. He did it. He won.

I threw my head back then, my eyes turned upwards to the sky. In the scattered blue light, I saw the sun on one side and the moon on the other; when I looked up, the heavens were mine.

We had ridden a beautiful race. If I had gone half a gear faster than Jibereen's cruising speed, he would not have had the finish in him to win the race. If I had gone slower than his cruising speed, he would not have picked up as well.

How did it all come back so naturally? Because it was always this way. Whether or not I wanted to be a jockey, riding was in my bloodstream; it flowed in my veins.

What nobody would know was that I had ridden this race using sound – my hearing – rather than sight – my vision – the whole way through. All of it, the whole thing, had been tactical, so as not to expose my weakness. It had been a risk of monumental proportions. But it had paid off.

Here's the thing about adversity.

It either rekindles hope or destroys it.

Which one wins? Hope or despair?

You have to believe it's hope.

The story of my comeback victory was on the BBC *Six O'Clock News*. Rarely is there a sports story, let alone a racing story, on the *Six O'Clock News*. But there it was. I knew then that I had achieved the impossible.

Everybody had thought me mad. 'Why would you even want to? Why would you want to put yourself through that?' they had asked.

Because I had to. The power of the human spirit is indomitable. What was for me, would not pass me by.

It hadn't been an easy journey. Mentally, it had taken an incredible strength of will to get myself back into the physical shape where I could even allow myself to think it possible, let alone then actually do it. And then, to win it, had demanded something more than I had asked of myself before. But when things hit you the hardest, you find something within you, an intangible force that propels you forward. You draw immense power from this inner strength, but many times you don't even know you have it until you are forced to depend on it.

I had to. So I did. And it was only belief that had made it possible. An unshakeable, unbreakable belief in myself. What we forget sometimes about belief is that it is self-fulfilling. If you believe you are capable of something, you give everything of yourself to live up to your own expectations. If you believe you are not, you give up before you even try.

I believed.

Chepstow celebrated.

Family cried.

Colleagues whooped.

Fans cheered.

Geoff Lewis eulogized.

Photographers clicked.

Journalists scribbled.

*

As for me, when I passed the winning post that day, walking up the dew-sparkled hill of Chepstow Racecourse, I felt many things. I felt happiness, yes. And pride. Satisfaction. Triumph. But more than anything – more than all of this put together, overshadowing every other emotion with a force so strong it tore down my walls – I felt relief. Just relief.

And as the tears streamed down, I didn't stop them. Because with the tears came the ultimate realization: I had played my part. I had repaid what I owed to the universe. I was free. Free from the burden of expectation. Free from the shackles of my mind.

I had done it – I had placed my flag on the top of the mountain.

There was just one last thing left to do.

My name was Declan Joseph Murphy. I needed to find myself.

And when I closed my eyes, I was once again that little boy astride Roger, riding free across the emerald fields of Ireland with my brothers and sisters, my head thrown back in laughter, my blue eyes shining with the sun.

Into the Lap of the Gods

'Sports people in general are a breed apart from most people.

But most sports people – footballers, golfers, cyclists – they are true "masters of their destiny", in one sense of the word, as to how much they want it; then it's mind over matter to go out and get it.

But as a jockey, you are actually putting your destiny on to a horse's back, you are offering it up. So yes, you are "master of your destiny" when you decide you want to do it, but you are still back into the lap of the gods where you started in the first place . . .'

Pat Murphy, brother, trainer and former jockey

Declan Murphy announced his retirement from the saddle shortly after riding and winning at Chepstow on 10 October 1995, only eighteen months after his accident.

His complete recovery from one of the most horrific accidents in sporting history is regarded by many as a 'medical marvel', and his comeback win can be considered as much a victory for him as a victory for medical science, in particular for Professor John Miles and his team.

Jibereen was the last win of his career.

Jibereen, Chepstow, 1995.

Epilogue

'Why now?' you ask. My story. 'Why twenty years later?'

It's an excellent question, one that I've asked myself so many times during the course of writing this book.

And the answer lies in the realization that in life, as in a story, timing is everything.

Timing changes the meaning of things.

Timing *gives* meaning to things.

Today, the only reminder of my accident is the ridged, crescent-shaped dent on the top right side of my forehead – the mark of the incision where my skull was cut open. My friends often joke about it when they introduce me to people who don't know who I am: 'Can't you see he rode horses,' they say, 'he's got a hoof-print on his head!'

I have been approached on three separate occasions with book offers, and once for a movie deal, but I couldn't bring myself to consider them. I have always been a deeply instinctive person and, somehow, something never quite felt right. But I do have my reasons, and I want to share them with you; perhaps you'd *like* to know, because by now, you have travelled this journey with me, alongside me, stride by stride.

First and most important, I am convinced that if I had shared my story immediately after my accident, racing would always have defined me and that's not how I want to be defined. I am a man, who happened to ride horses. But before all else, I am a man. I am just a man. My career was incidental.

Second, there are perhaps few people in the world more private than I am, and I never thought I could have shared what I had gone through with anybody – it seemed too personal, too invasive, too deep. I am, by nature, elusive, and there is an honesty demanded with a project such as this. An act of faith. I didn't think I had it in me, this ability to reach into the recesses of my heart, to bare my soul. Hand someone my life. What if they dropped it?

Third, what happened to me on 2 May 1994 was not one event, it was the beginning of a passage. A deeply personal passage. It changed me as a man – how could it have not? Fame is fleeting, memories last for ever, but when you're stripped of both, it really doesn't leave you with much. I carried a lot of pain inside me for a long, long time. But I didn't speak to anybody about it. Not anyone, not ever. I couldn't allow myself to. Our minds are flighty creatures, easily trapped; and the ability to be vulnerable is a gift – it releases the mind. In the eyes of the world, I may have recovered in a miraculously short amount of time, but when all that you have is taken from you, it takes a long time to heal. So for me to survive, I couldn't afford the luxury of vulnerability. I had to protect myself, from myself. And then, once I had got to where I wanted to get to, when I had achieved what I set out to achieve, it was over. Packaged and put away. A closed chapter, relegated to the dusty aisles of the archives of my mind – I didn't need to be reminded of it.

Until now.

I still cannot recall very much about the career that almost defined me. I have to be told which races I won, when I had won. But for the first time in twenty years, I have gone through newspaper clippings about myself, read the journalistic adulation for my rides, granted friends and family 'permission to speak', and I have been overwhelmed by it all. It was almost as if my life needed to advance away from it for me to be able to reflect upon it. Before the book, I would look back on my past with the detached curiosity of an interested third party. For the first time since the accident, I have allowed myself the privilege to cross that line. I have allowed myself access. It's OK, I've said to myself. It's OK.

It is therapeutic to set the mind free. When Ami and I watch my races together on YouTube, I note the steady progression in my riding: the Irish Champion Hurdle, the Queen Mother Champion Chase, they were good rides; the Mackeson Gold Cup was a great ride; the Tripleprint Gold Cup on Fragrant Dawn was an even better ride; and then, when I watch myself on Gale Again in the Silver Trophy Chase, just days before my accident, I feel like I am at the zenith of my career.

Could I have ridden a better race, or was that as good as it would ever get?

But this I would never know, because in the very next ride, it would all be taken from me.

And then, when I would eventually get it back – my life, my world, myself – everything else would have changed.

Perspectives change. People change. And so it was with me. When I came back, what I wanted from life changed. Because I changed. *The man who fell off the horse that day never really came back.* A different man did.

Sometimes, I wonder why I woke from that coma.

I wonder why I am still alive. *How* I am still alive.

When I look at it with hindsight, I think it almost impossible. When I was in the throes of it, I saw nothing as impossible.

'Do not go gentle into that good night,' wrote Dylan Thomas. 'Rage, rage against the dying of the light.'

How did I do this, how did I defy death? I don't know.

And this raises those difficult, and not always comfortable, questions of science and religion and plain dumb luck. Maybe it was my strength of mind. My will. That unshakeable grit that defined me. Or maybe it had nothing to do with that. Maybe it wasn't my time. Maybe Death visited me – in those days of unconscious darkness – and decided I wasn't ready to go. Maybe I politely told him to come back another time. Maybe he listened. Maybe he didn't. Maybe I just got lucky. Maybe . . .

But I did wake from that coma. How and why, I will never know, but I lived. And despite everything that was taken from me, it is the simple fact that I survived, that survives.

And so there was never a moment's anger from the day I regained consciousness.

Never a bitter moment.

Never why me.

Never ever.

I don't have a score to settle with life. I don't feel encumbered by it. I don't feel in debt. What happened to me was meant to happen.

Because at the end of it all, I consider myself incredibly fortunate. I am living a life that I may not have had. This is the song in my heart. Most people who suffer brain damage to the extent I did are not alive to tell their stories. And yet, here I am now – a walking, talking man; two eyes, two hands, two feet, and a brain. I can speak, I can hear, I can see, I can feel.

I can tell my story.

Everybody, from friends to family to people who barely knew me, universally said I changed after my accident. They said I lost myself in trying to become the man in the newspapers, in trying to become the person the world said I was. But what no one knows, until today, is that I couldn't remember who I was. I had no idea who I was. I was becoming who I was expected to be without any true understanding of who that person was. On the walls of my home hung the reminders of some former glory. Everywhere I turned, they were there. And yet I couldn't remember any of it. My life before my accident was like watching a movie, watching the sporting highlights of someone, once famous, now long gone, long forgotten. I cannot even begin to tell you how hard it is to reinvent yourself just to be accepted. So yes, I did change. I *had* to change. If I lost myself, if I became the man in the newspapers, it was because the newspapers were all I had to go by. My only sense of self was coming from what other people said about me.

No one knew this. There was no way they could.

Nobody sees the fight within you. They see what they see with their eyes; they don't see your spirit. On the outside I looked fine and nobody knew what was happening on the inside. No one knew the truth.

This book is my truth.

My first discussion of the book with Ami is a thirty-minute chat on a park bench. The next is at a Starbucks at Waterloo station, lasting a little more than an hour. I go into the first meeting convinced that I don't want to write a book; I come out of the second convinced that I do.

She asks me in the very beginning what the purpose of the book is for me; what is that one message that I want to leave our readers with. I tell her that it is simple. My purpose, my message, is Hope. I say to her, 'I want to tell people that what happened to me is not something to be afraid of. When I was in a coma, I was happy. It wasn't a bad place. I would have won either way. I was either going to wake up and be well again, or stay in the coma and be a child for eternity.'

When I say this, the tears well in her eyes. I crack a joke and she laughs. I know then that we'll get on. There's very few who laugh at my jokes . . .

Having someone write your story with you is not dissimilar to the act of riding a horse. In both cases, you are trusting another living being with your life. There are no broken bones, but there is all of the rest – dedication and courage and great skill. Sweat, blood, tears, passion, pain. Trust is everything. Mutual respect, paramount. My instinct is never wrong – it took under two hours for me to establish that we had all of this – an unspoken communication, an understanding, total and implicit. Our working dynamic has been the lifeblood of this book. Without it, *Centaur* could never have been.

But even then, and at several moments in the months to follow, I was totally unprepared for the full implications of the commitment I had made. I consider myself a fairly intelligent man – I take decisions after great thought, after weighing the odds, after considering all possible downside scenarios. Yet, there was so much that surprised me, so much that overwhelmed me. Neither did I anticipate how people who touched my life more than twenty years ago would come back into it to offer their help and support with such warmth and affection. Nor did I anticipate how emotionally charged it would be for me to relive those moments of my life through their eyes.

Little did I know that Colin Turner, friend and professional racing photographer, would not only send Ami the photos he had taken of me over the years but also enclose a handwritten letter with a moving account of my fall as he witnessed it – he was the person standing closest to the scene of the accident as it happened.

Little did I know that Joanna would be so forthcoming with her time, her friendship and her affection with regards to the book. Nor that her memory was so good – she even remembers the pub we went to on our first date (The Hole in the Wall in Great Wilbraham!) and, amazingly, what I ate (chicken and chips!).

Little did I know that Simon McNeill, Robbie Supple, William Jordan, Simon Claisse and Ian Bartlett would give us such a heartfelt reception at Cheltenham Racecourse when we visited it as part of our research for the book.

Little did I know that my old saddle still lived in the

weighing room there on Leighton Aspell's saddle rack.

Little did I know that Professor John Miles would say to me that he remembered 'every single detail' of my operation and would offer to share these with Ami and myself over lunch in the village in Wales, where he now lives a peaceful, retired life with his lovely wife, Enid.

Little did I know that Professor Paul Eldridge, Senior Consultant at The Walton Centre for Neurology and Neurosurgery, would give up several hours of his busy day to show us around the hospital, the wards and the ICU to allow us to recreate the hospital scenes with such authenticity.

Little did I know that the hospital administrators had a 'surprise' for me – in a small box, saved with great care for over twenty years, lay the breeches, riding boots and goggles that I had been wearing when I was brought to the hospital, unconscious.

Little did I know there was so much I didn't know.

My heartfelt thanks to all of the above named people. Your kindness humbles me. This book would not have been complete without you.

There are two people without whom – very simply – I wouldn't be alive: Professor John Miles and Joanna. My life is a testament to two things: one, your abilities to heal, and two, your faith in my will to rise.

There are many others who are vital to my story, and therefore to my life. In no particular order, I would like to express immense gratitude to:

Barney Curley, thank you for being the man behind it all.

Ross Campbell, for your friendship over the years and

for inspiring the weighing-room scene in this book.

Marten Julian, my agent who became an agent because I needed an agent. I am honoured and not in the least overwhelmed that you would dedicate your book *Strictly Classified* to me. You are absolutely right when you say in it that it was my spirit – and solely my spirit – that kept me alive, but of course, I am not surprised that you understand this.

The late Francis O'Callaghan for his irreplaceable mentorship at such an influential time of my life and for believing with such conviction that there was nothing I couldn't do.

The late Pat Eddery, for his invaluable advice on how best to ride Jibereen; I owe a big part of my winning come-back to your voice inside my head.

The late Josh Gifford, for giving me some of the best rides of my career and, more importantly, for being the man in the movie.

The late Jim Hogan, for his unwavering loyalty and all that it meant to me.

Yarmi, you were my greatest friend and my greatest frustration, both! But I live with the knowledge that you would have given your life for me and I don't know how one repays anyone for that.

Everyone who wished me well at the hospital and cheered me on at Chepstow, I thank you. Friendship is for all times; for offering hope in the bad times, for celebrating the good times – and you did both.

My brothers Pat and Eamon for giving so much of your time to help me revisit a very dark period in my life; of course, for your support, and for that of the rest of the Murphy family during that period, I will be forever grateful.

One of the most gifted riders I have ever had the pleasure of knowing, my sister Kathleen, whose close friendship in our youth fuelled many of the most endearing stories in this book.

My wife, Zulema, for being as rock solid as you are, and for the love and support you have shown, not only through this project, but through thick and thin. And for giving me our beautiful daughter, no thanks will ever be enough.

My daughter, Sienna. When I sometimes wonder why I woke from my coma, I look at you and I know the answer. You fill me with joy, you fill me with happiness, you complete my life.

Ami, in telling my story, you have set me free.

I feel alive to the fabric and the texture, the weight and the wonder of my life.

I want to be the person in this book. I want to be me. I feel proud to be me.

It has taken courage to be this honest and there is nothing more uplifting than having the freedom to say this openly.

Still, it hasn't been without its fair share of drama. Given my memory loss, Ami and I have had to piece my story together like some elaborate Cubist construction, bit by bit, slowly and painstakingly, the scattered pieces of a giant jigsaw that make up the life of a man; a process that has been heartbreaking, heart-warming and liberating, all at the same time.

The first time I tell her about the lost years, she questions how this book can even be written. 'How do you possibly tell a story you can't remember?' she asks. Quite rightly.

A few days later, she is over her private moment of panic. 'It's not going to be easy,' she says. 'The book. It doesn't mean we won't try to make it amazing, but it won't be easy. It might hurt. Actually, to make it good, it'll *have* to hurt. We're going to need to compensate. I need to know everything about you that you can remember. Everything. What makes you tick, what keeps you up at night, the crimes you have committed, the songs you sing in the shower, your pleasures, your pain, your unrequited loves, your greatest insecurities, your memories, your music, your mind . . .'

And so the same person who thought his story was too private to talk about, tells her whatever she wants to be told. I tell her everything.

Later, I say, 'I have cried more tears because of you and this book than I ever imagined crying, for anything, in my entire life.'

She says, 'That, there, just made this whole thing worth it.'

Ami's Note

In the orange October of 2015, some mad urge possessed me to make the most reckless of claims. I said to Declan Murphy, 'I could write your book.'

There was nothing we had in common, save a fascination for the same city, a love of horses, a penchant for words and a zest for life. Which made me about as qualified to write the life story of a former jockey as a penguin is qualified to write on the mechanics of flight.

And so – more importantly – when I wonder what mad urge possessed *him* to take that chance on me, I don't have an answer. Fate or Providence, I will never know.

But Declan's life is a triumph of human resilience. And this is a story that needed to be told. That much I have always known.

What I could never have known is that in telling it, I would unknowingly gain so much.

There is a line in the book that stands out in my mind: 'We are capable of infinitely more than we think.'

Many people have remarked to me over the course of this past year that *Centaur* mustn't have been an easy book to write. It wasn't. But this had less to do with my lack of familiarity with the subject matter and more, much more, to do with how much of myself I needed to give, to feel what I needed to feel to be able to write this story. The authenticity of the emotion that runs through the book is the pulse of *Centaur.*

To write the kind of book we wanted to write, we had to go back many times into the depths of darkness, to that place where hope almost died. To see a grown man – a strong, proud, grown man – cry openly in front of you is a life experience. And yet we shied away from nothing.

I applaud my subject for his courage. Honesty takes a whole lot of courage.

Indeed, in the spirit of honesty and to fully appreciate *how* this book was written, it is necessary to mention Declan's memory loss. There was a consequence of bringing the knife to the brain. A life had been saved, but there was a price to pay. There are years missing, significant ones; ones that would prove crucial to the telling of his story. But – together – we soldiered on, despite it. And perhaps with stronger resolve, *because* of it. Time lost and time regained have been some of the most complex and fascinating concepts that *Centaur* explores.

Accordingly, the portrayals of the races in this book – or, for that matter, most of the goings-on in Declan Murphy's world between October 1989 and May 1994 – are not just generic recountings of how a sportsman remembers his career. Instead, they have been put together, stride by stride, fence by fence, through the meticulous research of archives, endless readings of news clippings, repeated viewings of YouTube videos, and countless conversations with people, over and over and over again. The result, ultimately, is the story of his life, coloured by my imagination; an intricate reconstructed collage where real things are embedded within, and enhanced by, the fiction of the mind. We do this because of our circumstances, because we *need* the beauty of multiple perspectives on the many unanswered whats, whys, whos and whens, floating like snowflakes through this spectacular deception of Time. For me, never has any endeavour proved more challenging – and equally, more joyful.

My biggest rewards in writing this book are the people I have met from previously unknown worlds, who have taught me that learning never stops. The world of racing is as fascinating as it is different from the world of medicine – in assisting with my research, I would like to thank, in equal part, Declan Murphy's racing friends, fans and associates; and the exceptionally talented team at The Walton Centre. I stand in awe and have such deep respect for how sanguine people are in these lines of work. The universal willingness to help, just for helping's sake, says something really positive about humanity.

My special thanks go to Declan's warm-hearted, fun-loving and authentically Irish family – we met as strangers and ended up as friends – thank you for opening your homes and hearts to me.

Jo – meeting you has been serendipitous in so many ways. You were my most unexpected treat, a true kindred spirit. How can I ever thank you enough for giving us 'Joanna'.

If not for the two Tims – Tim Hayward and Tim Bates – *Centaur* could not have been brought to life.

Tim Hayward – writer, broadcaster, restaurateur – my

gratitude to you knows no bounds. Thank you for your belief in my pen. And thank you for 'paying it forward'.

Tim Bates, you are indeed, in Tim Hayward's words, a 'cracking good agent'. We wouldn't be here if not for your smarts and your guidance in championing our cause – I can't believe how lucky Declan and I are to have found you.

We can never fully express our appreciation to the truly exceptional team at Transworld for your generosity, conviction and confidence in this story. A sea of gratitude to our editor Giles Elliott, for your unchecked enthusiasm and true understanding of what we were setting out to do. But mostly, for inhabiting our book. We've learnt that it takes Herculean behind-the-scenes efforts to get from words on a laptop to a real, live book. Vivien Thompson, Helena Gonda and Leon Dufour, please take a bow. Rebecca Wright, the cleverest copy-editor on the island, without you this book would be longer, 'wronger' . . . and worse. Every writer needs an Alison Barrow on their side and we were fortunate enough to have her on ours – your vision has allowed *Centaur* the opportunity to reach every reader it was written for.

My deepest thanks to my family and close friends for your encouragement, advice and good sense. Especially to my mother, for reading countless iterations of every chapter with such a keen and insightful eye. My boys, Ranbir and Reyaan, for your infectious exuberance for everything I do – every girl needs her cheerleaders.

I have only been able to write *Centaur* in the timeframe it was written because of my husband, Sid: my first reader, my toughest critic and my greatest strength – without your support neither I nor this book would be.

At the end, I need to go back to the beginning, and the story of *Centaur* begins with the story of one extraordinary man. The enormity of shouldering the responsibility of writing someone's life story – especially one where chunks are lost – had me shrouded in a smog of self-doubt so many times. And yet, the person who had most to lose from my mistakes never once doubted me or my ability. He *believed*. Declan's positivity – idealism, even – with this book has given me limitless power as a writer.

Declan Murphy – your life is your greatest celebration; your story is mine. Thank you for trusting me with both.

We are capable of infinitely more than we think.

Declan Murphy was born in rural Limerick on 5 March 1966. Like most of his seven siblings, he took to riding horses from an early age and after being spotted by Ireland's top trainers became a leading amateur jockey while at school. He then moved to England and rode a host of winners in races as prestigious as the Champion Chase and Mackeson Gold Cup, as well as two Irish Champion Hurdles, before a near-fatal accident on Arcot at Haydock Park in May 1994. Eighteen months later he rode a final winner, Jibereen, at Chepstow.

Ami Rao is a British-American writer who was born in Calcutta and has lived and worked in New York City, London, Paris, San Francisco and Los Angeles. Ami has a BA in English Literature from Ohio Wesleyan University and an MBA from Harvard Business School. As a self-proclaimed foodie, she has written before for Tim Hayward's *Fire & Knives*. She has always been 'absolutely horse mad' and rides regularly in her spare time. *Centaur* is her debut book.